Even The Caterpillar Sings

Bringing Our Souls Back Into A Deeper Relationship With Nature

iRewild Institute
Ocala, FL

Written by

Leah Black
Ida Covi
Louise Livingstone
Larry Pappas

Illustrated by

Daniel Desembrana

Edited by

Hannah Irish

For our world . . .
a world in which we all belong

This is an iRewild Institute Book
Ocala, Florida
www.iRewild.com
iRewild Institute is an international organization working to bring the human soul back into a deeper, more conscious relationship with nature. For information about how you may participate in its programs, please visit our website.

Copyright © 2024 by iRewild Institute
Text copyright © iRewild Institute
Illustrations copyright © iRewild Institute
iRewild is a registered trademark of iRewild Institute

Written By Leah Black, Ida Covi, Louise Livingstone, Larry Pappas (alphabetical order)
Illustrated by Daniel Desembrana

All rights reserved. No part of this publication may be reproduced, stored in a retrieval system, or transmitted, in any form or by any means, electronic, mechanical, photocopying, recording or otherwise without the prior permission in writing from iRewild Institute. Nor may the pages be applied to any materials, cut, trimmed, or sized to alter the existing trim sizes, matted, or framed with the intent to create other products for sale or resale or profit in any manner whatsoever, without prior permission in writing from iRewild Institute.

The characters and events portrayed in this book are fictitious or are used fictitiously. Any similarity to real persons or characters, living or dead, is purely coincidental and not intended by iRewild Institute, the authors, or illustrator.

Published by

iRewild Institute, Ocala, FL, USA
and
Certeza & Sons Publishing, Laguna, Philippines

iRewild ISBN: 9798870072661

Library of Congress Cataloging in Publication Data

iRewild Institute
Even The Caterpillar Sings
Bringing Our Souls Back Into A Deeper Relationship With Nature

To Nature

Whose infinite beauty and Earth-born
companions call out to our souls, inviting us
to experience the deepest realms
of human existence.

Introduction

We sometimes experience moments in life when time seems to stand still—moments when we find ourselves spellbound, and eventually transformed. These moments can be cosmic in magnitude, as reflected in the awe we feel when we see the night sky brimming with stars and realizing that there may be 2 trillion galaxies in our universe. Or these instances may be inwardly profound, as when we witness a baby animal frolicking with its mother or stand before the grandeur of a 200-year-old tree. These experiences seem to burrow themselves into the innermost depths of our beings. They move our hearts, evoking experiences that lie beyond words. In Nature, we encounter a feeling that transcends our individual sense of self and exceeds the boundaries of the intellect—a feeling of what is infinite, divine, sacred, or god-like.

When we open ourselves up to Nature, when we see our world as a creative, innovative, living presence, we can heal the fracture between humans and the Earth. We commingle in a reality in which there is no separation between the physical and the spiritual, between mind and matter, between heart and head, or between self and an-Other. We find ourselves, if even for a few moments, participating within the deepest realms of human existence.

Inspired by the breathtaking beauty and wisdom of our planet, Even the Caterpillar Sings is a collection of reflections that unlocks a door for your heart and imagination to rediscover your intimate connection with Earth and renew your reverence for our natural world. By nurturing a relationship with Nature, you are also nurturing your own being because you and Nature are inseparable. Every time you engage with Nature, you connect to your innermost essence, you revive your senses, restore your overall health and wellbeing, nourish the soul of our planet, and help create a world in which we all belong.

Week 1

Imagine

Take your mind back 7,000 years ago, to when people lived in wild lands, untouched by the hand of modernity. Humans lived in union and reciprocity with Nature, fulfivlled and thriving with all we needed to survive and to grow physically, emotionally, mentally, and spiritually. Our ceremonies, prayers, dances, and rituals were first inspired by the whisper of healing plants, wise animals, or sacred winds. Nature was our teacher, family, and co-creative partner.

These times of peace between Nature and people are now mostly utopian memory. Times when we once left only a fleeting footprint upon the landscape. An era when we shared our love for Nature by celebrating and honoring her cycles of life alongside our own. A world that lived to its full potential around us, blanketed in Nature's wilderness, wisdom, biodiversity, and abundance. Just imagine the piercing sounds of wild, untouched rainforests, jungles, woodlands, grasslands—oh, how the crickets and birds must have sung! Can you even conceive the scent of a thousand wildflowers, in a single aurochs-mown glade, filled with the buzzing ensemble of bees? It feels barely imaginable, yet our minds drift back like it's somehow familiar terrain, written, maybe, in our DNA, or stored in the core memory of our souls and Earth's soul, intertwined.

Question

Over time, Nature has been controlled, reduced, cut back, and restrained by humans. If you look at yourself, as Nature, deep within yourself, and outside of yourself, exploring the expanse of your personal ecosystem, what has society preened away and tamed from your original being? How would it feel to re-wild yourself, to thrive once again like an untouched forest that's been hidden away, safe from manipulation, in an unreached volcanic crater? How can you re-wild yourself back to who you truly are? Enable yourself, and your soul, to release your being, your roots, your branches, your body, to live to your fullest wild potential.

Explore

Find a place where you can clearly see the horizon, maybe out of your window, or in an expansive rural area. Watch the landscape, as far as your eyes can see. Allow your mind to be mesmerized by the horizon. Pay attention to all that is in front of you, and around you, up towards the ever-expanding boundary of Nature's scope. A horizon that could be, literally, anywhere on Earth right now, as her ending expands and curves on a continuous cycle, like life itself. Maybe the horizon is at the edge of rolling pastoral fields or vertical unconquered cliffs, meandering roaring rivers, a dark beckoning ocean, or at the end of the bright city lights.

Give your imagination permission to form a dynamic image of the landscape surrounding you as it would likely have been 7,000 years ago. In your mind's eye, envision the life and land around you, changing and re-wilding back to how it naturally was. Open your imaginative senses and smell the air of 7,000 years ago; listen for sounds of mammals, birds, and insects; hear the activity of humans and wildlife on Nature's land. What do you see? What elements do you sense? What wonderful plants and trees lace the land around you? What scents do you breathe into your body? How dense is the vegetation? How different does it look and feel? How do you feel?

Our ancestors of 7,000 years ago created magnificent artwork on cave walls of bygone communities that we admire in all their wonder today. 7,000 years seems so far away, yet it is only a rough stepping stone in the existence and journey of humanity. Still today some people call caves their beloved homes. If you feel drawn to, honor your ancestral land through a piece of art or creative writing, like your ancestors may have done. If you'd like, you could offer your piece of creativity back to the land you stand upon, to show Nature that you remember her, honor her, and are grateful for her never-ending support. Your love has the potential to heal and re-wild the world.

Walk now through the land of your ancestors, of your home, imagining it as it used to be: wild, bountiful, beautiful, and giving without wilting and waning. During this walk, acknowledge your relationship with the word "wild." Do you see "wild" as it is sometimes described in dictionaries of modern times (uncivilized, savage, uncultivated, beastly, feral, untamed, uncivil, undomesticated, barbarous), or do you see it as cherished family, like many of our ancestors did? Maybe you see "wild" as something else. What does "wild" mean to you?

iRewild
Can you think of any piece of land on Earth that is not "owned," in one way or another, by humans? Can you think of any species on Earth that takes "ownership" of land and entire

habitats, ecosystems, countries, and continents like humans do? Think back to your recent ancestors and their possible ownership and colonization of land, people, and wildlife. In acknowledgement and gratitude for all Nature has done, even sacrificed, willingly or unwillingly, for you, your family, and your ancestors, re-wild a portion of land you own or rent, even if that's simply a windowsill planter box. Nature will thank you for the smallest act of love.

Imagine the scope of change, if every person gave back to Nature a portion of "their" land, to help create a planet where Nature is given freedom to re-wild, to exist. When you become the ancestor of future generations who will enjoy the beauty and biodiversity of wild Nature, would that not feel like a life well lived? You have the chance to help the world through re-wilding yourself and Nature. Even a small piece of land given back to Nature, per person, could grow into a great slice of wild, rekindling and replenishing our relationship with Mother Earth.

Reflection

Week 2

Imagine

In the natural world, time slows down. Plants, wild animals, and trees move with the ebb and flow of the cycles of life. Autumn winds coax golden, red, and copper-colored leaves from the branches where they lived throughout the spring and summer. The leaves gently waft and spiral to the forest floor, softly settling to provide nutrients for the soil, and food and protection for the myriad of tiny creatures who live on the ground. Wild animals forage and sniff the air, weaving back and forth, taking time to simply be in the world in which they live. Plants take time to come to their full maturity, unfurling into their authentic selves with quiet, slow grace.

Question

Are you fully experiencing the natural world? Nature may be the place where you exercise or simply pass through without mindful attention. What if Nature is inviting you to move into her own time and pace? What if you moved into the possibility that Nature is inviting you into a different experience?

Explore

Stand under an autumn leaf shower. If it's not autumn where you are, imagine that you're standing under a myriad of leaves that are spiraling and whirling all around you, being wafted by chill autumn winds. Or, over the course of several months, you could spend time in Nature visiting a plant or a tree that particularly speaks to you.

iRewild

Contemplate this statement: "What is this life, if, full of care,/ we have no time to stand and stare?"[1] Having completed the activities above, has anything changed in your life? If so, how, and how might you practice your new-found understanding in the world?

---

1   W.H. Davies, "Leisure," in The Routledge History of Literature in English—Britain and Ireland. 2nd ed. Ed. Ronald Carter and John McRae, (London: Routledge, 2001), 308.

Reflection

Week 3

Imagine
Imagine you are a groundhog, standing on the hillside, enjoying the sun and the blue sky dotted with clouds. Calm, peaceful, and untroubled, your hands gently reach for the top of a tall flower. What brings you out from your earthen burrow, this opening from one world to another?

Hidden beneath the landscape, your ground engineering skills are critical—yet, so easily overlooked and unappreciated. Your tunnels ensure plant roots receive their necessary oxygen. Your digging turns the soil, mixing nutrients into the dirt to feed other organisms. Your elaborate, interconnected tunnels provide shelter for many animals, and you generously share your home with fellow hibernators during the cold winter months. Your work is connected to something beyond you, to something infinite.

Question
Our relationship with Nature is swayed by our concept of what is, in fact, "Nature." It's driven by our beliefs surrounding how we see ourselves in comparison to Nature, and it's

influenced by our actions and reactions toward the natural world. How do you define your personal beliefs about Nature and your behaviors toward our natural world? Explain why you feel this way.

Do you see yourself as separate from Nature, something that exists "out there"? When do you see Nature as something humans should control? When do you believe that Earth's natural resources should be utilized, without constraints, to meet our needs?

Do you assume Mother Earth can take care of herself? Alternately, do you see Nature as a source of life to be respected?

When have you experienced the sacred in Nature? When have you experienced Nature being infused with consciousness and possessing creative intelligence? Or, when have you experienced the natural world as an interconnected, interdependent part of the whole, in which humanity is just one part, neither more nor less than the rest?

Explore
Take a few deep breaths and look around. Are you in the city, surrounded by concrete and street lights? Are you sitting outside in the wilderness, an urban park, or garden? Are you on a bus, train, or in a car? Are you in your home or at work? For one brief moment, regardless of where you are on this Earth, consider that Earth is providing you with everything you need to sustain your life. At every moment, without reservation, Earth is giving you air to breathe, water to drink, food to eat, and the necessary resources to make your clothes and build your home. Earth is not a lifeless place where we live out our lives. Mother Earth takes care of you, your children, family, and pets with pure, unconditional love.

Close your eyes and take a deep breath. As you breathe in, experience and focus your awareness on Earth's life-giving gifts to you. Surrender to your heart, surrender to Earth's unconditional love and allow it to carry you into a place of unity with all of Nature. As you inhale, welcome each breath, admire the clouds in the sky, enjoy the birdsongs, appreciate the ocean waters, savor the taste of fresh berries, bask in the sun's rays, and celebrate the wonder of trees as they provide you with another breath. Give yourself some time to soak in all that you have experienced. See Earth's love as you breathe in understanding. Awaken to feelings of deep respect and love for the beauty and generosity of Earth.

iRewild

A meaningful, deeper human experience often comes with caring for an-Other. As we nurture our love for Earth and our passion for our common home alongside Nature, we respond with care and responsibility. All our lives are sustained by reciprocity.

How can you nurture a moral principle of reciprocity for Mother Earth within yourself, your family, and your friends? Rethink your daily habits so that you can give back, so you can participate in a mutually beneficial relationship with Nature. What gifts can you offer to Nature so that the living Earth knows that her work of caring for all life has been seen, valued, and respected?

Reflection

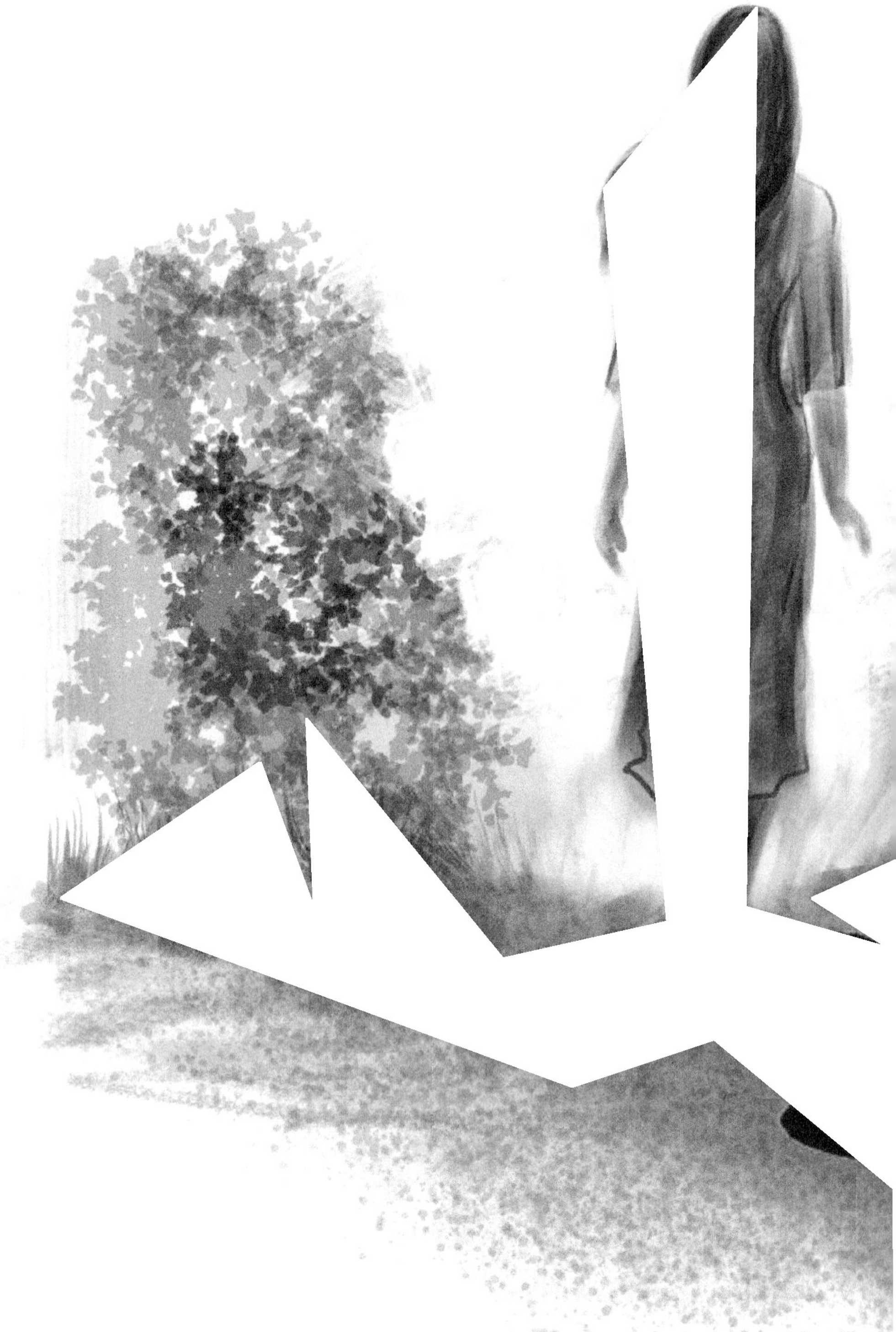

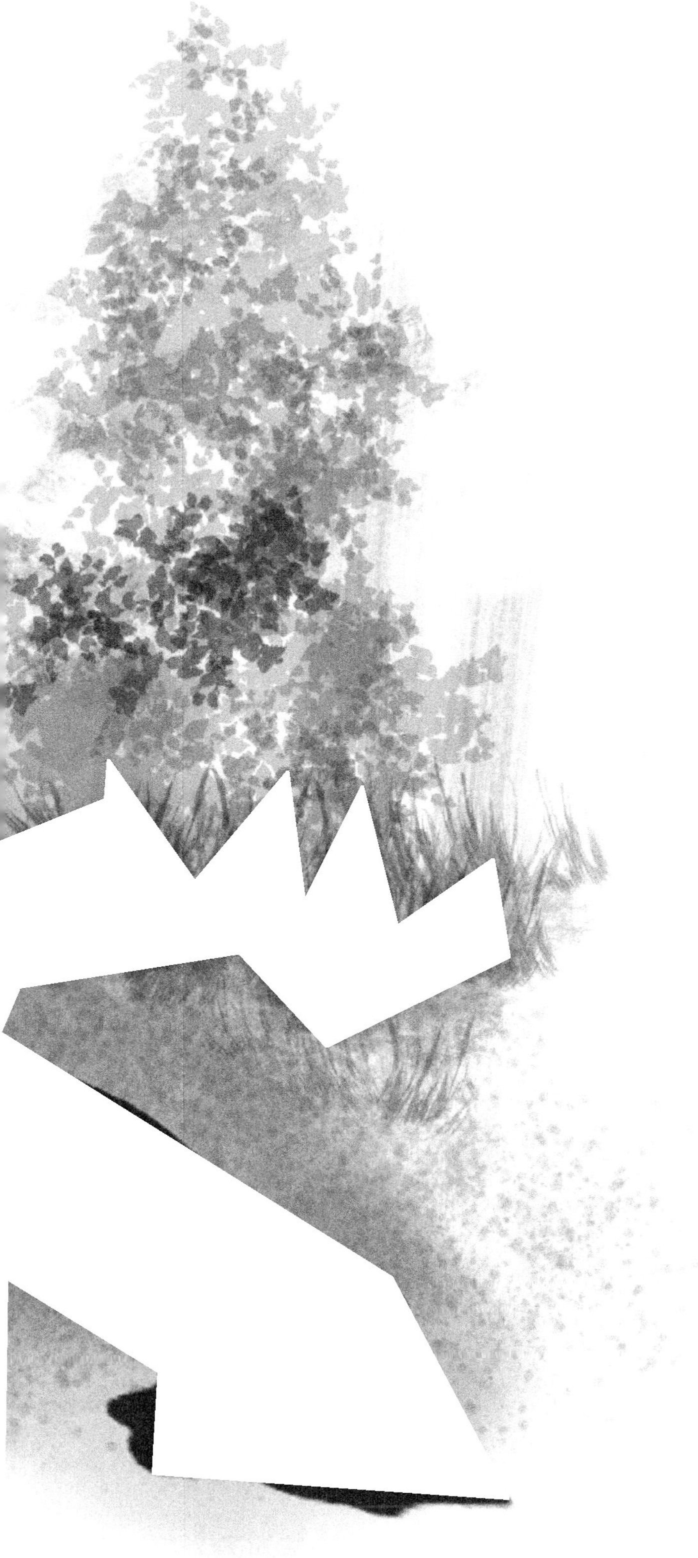

Week 4

Imagine

Seeing a bear in the deep wilderness is a reminder that you are a guest in her home, her natural habitat. As you meet, she quietly runs off, and yet, you are left trembling, setting your inner life in motion. The bear awakened cultural beliefs that you were born into, which secretly influence and shape your thoughts, actions, and interactions. Within your inner world, within your stories, you become entangled with the bear, a symbol that embodies aggression, a danger that is always lurking, and a dominance that renders you powerless. Imagination is powerful. However, if you look closely, it's not the bear you fear.

Because of certain cultural beliefs, bears are disappearing from our world—the consequence of how myths manifest and lurk in the shadows of our unconscious, individually and collectively. Why do we attribute such qualities to her, this beautiful being? You ask the bear for forgiveness, because, in actuality, her spirit is curious, playful, nurturing, brave, and protective of her young.

Question

Are you able to imagine yourself walking by the bear's side? What cultural beliefs and qualities do you attribute to wild animals?

Explore

Remember a time in your life when you encountered or held a special kinship with an animal in the wild. What does this animal represent to you? Research the characteristics, qualities, and behaviors of this animal. Since Nature is highly interconnected, how does this animal help to maintain its surrounding ecosystems?

Look for a place where you can go to observe and appreciate this animal, even if this is your imagination. What can the spirit of this animal teach you? Thank the animal for its purpose here on Earth, for its meaning may be far greater than we have ever imagined.

iRewild

How can you make sure your children, friends, and family develop an awareness and understanding of Nature to see animals in the wild for who they truly are and the importance of protecting, and living in harmony with, the planet's ecosystems?

Reflection

Week 5

Imagine

As you care for all the animals on the farm, you gradually develop a profound understanding of their world, enabling you to break free from any preconceived biases. You move closer to observe and appreciate every animal with deep respect and admiration. Through a fresh perspective devoid of prejudice, you are profoundly touched by the richness and diversity of their intellect. Each one is unique. Each one has its own personality. Similar to yourself, they all have their likes and dislikes. Some are bold, exploratory, and social, whereas others are quiet or solitary. Many animals display curiosity and trust, whereas a few adopt a more distant and observant attitude. The majority are easygoing, but a few are pensive. Certain ones embrace change, while others show absolutely no interest in it. Most of them are extremely imaginative, while others prefer the ordinary. Every so often you see them play and frolic, truly enjoying having fun. You see them express love, form strong bonds, and nurture their young. From time to time you see them staring at the surrounding landscape as if contemplating Nature's beauty or, possibly, life. They experience profound pain when injured, grieve the loss and separation from their loved ones, and feel fear when facing death. They cherish their lives and share the same desire to live as we do. This serves as a profound reminder of the inherent intelligence, preciousness, and irreplaceability of every single life.

The experience of consciousness is universal among us. We all share the gifts of intelligence, emotions, creativity, pain, joy, and sorrow. Our diversity is woven together in a complex interconnectedness, forming the fabric of our existence. Despite our differences, a unifying thread weaves through all beings—the universal longing to thrive and experience the essence of being alive.

Question
We are born into a culture where we live by a gallery of beliefs that are anchored deeply in our inner worlds. In what ways do you view plants and animals as limited just because their intelligence does not reflect your own? Can you perform photosynthesis like a tree, converting sunlight into oxygen for living beings on Earth? Where do you draw the line?

Explore
Take a moment to witness the wonder of a spider weaving his spiral web from fine silver threads strong enough to stop the high-speed flight of an insect. Observe ants as they work together to carry away organic waste, contributing to a cleaner ecosystem. Listen to the symphony of flowers as they sway in the wind attracting bees, ensuring the creation of fruit and seeds. Pause to appreciate, value, and express gratitude for the many diverse forms of life that surround you. Each being serves a unique purpose and supports the interconnectedness of existence, including your own.

The next time you venture into the forest, take a moment to seek permission from the beings that call that place home. Remember, you are a visitor in their habitat. Before departing, offer your sincere gratitude as you would to any gracious host. Similarly, when encountering a tree, weed, butterfly, fish, or any other living being, pause to acknowledge and express appreciation for the essential role they play in forming and maintaining the delicate balance of the biosphere. Each being contributes to the vibrant tapestry of life that sustains us all.

While we may never fully know an-Other being's experience of joy, fear, or sorrow, we can still show respect and mindfulness towards them. The next time you pick a flower, harvest vegetables, or cut down a tree, take a moment to communicate your intentions to the plant. This allows the plant to withdraw its energy from that area, share its knowledge with others, and potentially reduce any pain it may experience.

iRewild

Drawing from your personal experiences, what steps can you take to embrace a deeper awareness that we coexist in an animate world where every being possesses consciousness, intelligence, and the ability to feel? What actions can you take to reinforce this awareness and contribute to the well-being of our living planet?

Reflection

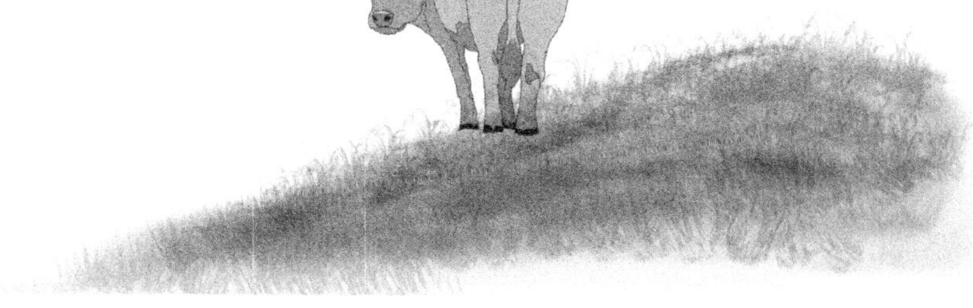

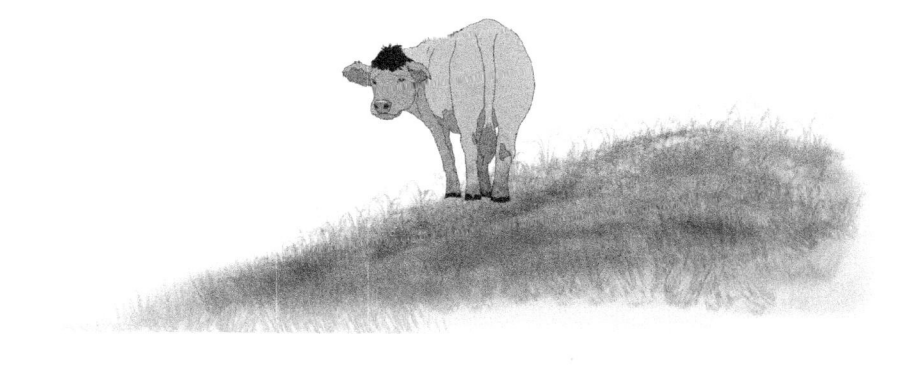

Week 6

Imagine
You are standing on a white-pink sandy beach looking out across a crystal-clear azure sea. As you wiggle your toes in the soft, powdery sand, in the far distance you can see black fins cutting a pathway through the water. You wonder, do these fins belong to a dolphin, a whale, or a shark? Perhaps you recall stories of people who have swum with sharks and experienced a deep, joyful connection with these animals. Or perhaps you feel a slight tingle of nervousness. You are always in relationship with the world, just as the world is in relationship with you. How might the shark see you, sense you, as you stand on the beach staring out into the inviting sea that these beautiful creatures call home?

Question
At what moment does a graceful, powerful shark become a predator? Before you answer, take some time to breathe into your heart. Looking from a different perspective, how does the shark feel when you enter his own domain? What rules apply there that you might have overlooked? Look more deeply at the world around you. Can you allow yourself to consider the moment that sharks become enemies? How does your framing of the world create the world that you see around you? Is the world really the way we have been conditioned to believe? How might your view of the world shift through the eyes of the shark?

Explore

Close your eyes and breathe slowly, in and out, through your heart. If you have difficulty with this, bring something into your awareness that warms your heart—someone you love, a beloved pet, a landscape. As you breathe, flood your body with the heart's benevolent qualities of love, compassion, kindness, empathy, openness. Now move slowly out into the world. As you step outside, make a commitment to meet the world as though you are seeing it for the first time, as with the eyes and innocence of a young shark.

iRewild

Consider this statement: We are always in a living relationship with sharks and the world. Within this connection, there is no hierarchy, simply two beings in union endeavoring to really hear one another and find ways of acting that honor the needs of both. How can you bring this perspective into the world you inhabit and the watery worlds into which you immerse yourself as a guest?

Reflection

Week 7

Imagine

The world of soul is invisible. The Greek philosopher, Plato, was the first to describe the world as a living thing, "truly endowed with soul and intelligence."[2] In our daily life, we often talk about things with utmost certainty. Yet, our universe extends beyond what we can perceive with our senses. Its mysterious, underlying reality is often hidden from us, yet its unseen existence is essential for giving rise to things that we sense. The hidden is as real as the visible.

Physics tells us that time may be an illusion onto which we attach our past, present, and future experiences. Yet, time governs our lives as it drains away our days, months, and years. Consciousness is hidden from sight, yet this thing, process, or energy allows us to experience life. Thoughts are completely invisible, and yet they influence every decision and action we pursue or abandon. Imagination cannot be detected, touched, or seen, and yet it is the source of all creativity, ideas, and innovations, and is a facilitator for interacting with the world around us. Air is all around us, intimate and faceless, and each breath of air provides us life.

The silence of time, consciousness, thoughts, imagination, and air were at one time mystifying. Each of them keeping to itself, each of them challenging our commonsense notion of what we see. We are limited beings participating in a world where Nature is busy at work beneath the surface, where we encounter the generosity of invisible elements ruled by invisible truths. We live between the known and unknown. Between the visible and invisible. Between the revealed and unrevealed. Between darkness and light.

---

2   Benjamin Jowett, "Timaeus by Plato," The Internet Classics Archive, https://classics.mit.edu/Plato/timaeus.html, accessed on August 21, 2023.

Nature can only be understood when we reach beyond our typical ways of looking at things. To experience, to experience the depths of existence, to experience things that are hidden from sight, we must transcend the observable realm of space and time through our imaginations and with our hearts. The life essence of Earth's soul is also rooted in the human imagination and embodied within our souls. Imagination and heart are our inner senses, always lured to the hidden, coaxing the intangible to emerge. We create a bridge from our rational mind to our inner world, and then to the soul of the outer world.

Though completely invisible, soul is the innermost essence, an underlying ingredient, that allows for all things to have an internal dimension; no matter what shape it takes, no matter how simple or complex a system it is, it still has a hidden inner world. Through soul, all of life, including our living, breathing Earth, takes on a form, an essence, and enters our awareness. Every life-sustaining tree, every resilient stone, every vitalizing drop of water, is alive. When we meet things on a soul level, the source of life itself, we illuminate our consciousness. Boundaries wash away and the whole world comes alive. It makes meaning possible, and gives everyday experiences a fuller, richer, deeper significance. Soul then moves from being a single possession for humans alone to being collective. Our awareness shifts from just being viewed as my soul, my fate, my biases to being interwoven within a living world, a larger soul, that encompasses our natural world. Nature now becomes an extension of ourselves. We awaken the rigidity of our minds to consider, to discover, that everything in our natural world is permeated with life, intelligence, and soul. Isn't the mysterious invisible beautiful?

Question

Do you perceive and value things by only their outer natures? In what ways do you consider their inner natures, their soul qualities, the inner light they bring into your world?

Explore

In the early morning hours, step outside to meet the sunrise, or open a window where you can see the soft, colorful light of daybreak. Take off your shoes and feel your feet on Earth's soil. Feel the fresh morning breeze as it softly touches your face, arms, and feet. Recognize this breeze as Mother Earth gently touching you with her soul. Breathe. Acknowledge the interflow between you and Nature, between you and the soul of the world. With each breath, Earth's soul provides you with life, freely sharing its existence, its beingness, with you.

Visualize the golden rays of morning's early light stretching across the landscape as the soul of the world. See yourself through the eyes of Earth's soul, revealing her true self to you, awakening you with limitless love, embodying you with life, lighting a path home, accepting your weaknesses, nourishing your heart, healing all worries, calling all the pieces of your lost soul back to you, and inviting you into a unified world. Immerse yourself in the visualization. Without asking why, fall into, merge with, experience the soul of the world as it comes into full union with your soul, extending everywhere, touching the stillness of your inner world, touching the landforms of your outer world, sharing itself unconditionally. It is an unquestioned exchange of existence with the divine oneness of Nature—the whole unifying all life, becoming one, our destinies entwined. With awe and humility, use all your senses to experience this bond with the living fabric of Nature, the living essence, the beingness of the soul of the world.

iRewild

How does the soul of the world see and sense you? Explain what you are feeling—the emotions, the sensations, the wonderings, the future stories. Describe everything you observed, everything you felt, in intimate detail. Draw a picture that describes how Earth's soul sees you and share your drawing with Earth's soul as a gift of your deep connection. Let the light shine through your drawing in a place that has a special meaning for you.

Reflection

Week 8

Imagine
The far side of the moon—always there, looking out into space, but never at us. He circles us, creates a space between us, his back to us, allowing us to only imagine him from what is revealed to us by his opposite. He is satisfied with looking away, protecting us from the unknown, from the rocky, metallic, icy messengers from afar. He is our sentry, always on guard, humble, yet proud. He doesn't tire of his place or his purpose. He, too, can only imagine us, never knowing us. We give him a purpose, a life of generosity, without wanting anything from us. Each of us, never knowing the other, revolves around the other, imagining, wondering, what may be for the other. We sense the sun, the Earth beneath us, other beings. We connect with all that we sense with intention, proximity, representation, even depiction. But the far side of the moon stays away, leaving us to imagine.

Question
Connecting can be perplexing for many of us. At times, it's easy to form a bond with a new person in our lives, with a beautiful vista, or a loving animal. At other times, we find it difficult to reach a person close to us, in our family, or a life partner. Our hearts and minds can take us to places we are not able to comprehend, to an understanding without

really knowing. They can also confound us at times by tugging us in an unusual direction. What about things we have no immediate relationship with? Or almost no knowledge of? Is deep connection possible? Much of our natural world is beyond our physical grasp, merely in our minds, maybe our hearts. And some aspects are even beyond our thoughts, unknown in existence, even in our imaginations.

Can we have a conscious relationship with that which is beyond our senses, our thoughts, our imaginations? An ability to emotionally and consciously connect with far away aspects of Nature may help us, and be the key, to bond with that which is close to us, but remains elusive. There is an ease, a freedom that comes with the clean newness of an experience. Can we extend this ability, this gift, to developing a conscious relationship with the unknown, the depth of Nature we do not even contemplate? If we are to develop an understanding and bond with the sacred in our immediate natural world, might the free, imaginative exercise of joining with the unknown aspects of our world, our universe, cast off any reluctance, difficulty, or non-acceptance associated with recognizing the soul of the world in the simple sapling, the unassuming stone, and the quiet stillness of a breezeless morning?

Explore

Day or night, visible or not, the far side of the moon is there. An infinite number of things are beyond, unseen, unheard. Some we know well through learning even without seeing. Some we barely know either through limited contact or study. Still others we know nothing other than their names as we have given them. Lastly, there are those we know nothing about, not even a name, or even their existence. How can we connect to something we have not, and may not ever, see, touch, hear, know, imagine? Step outside on a moonless night. Try to contemplate the existence of all that is unseen in your world, those things that you know of that may be there, the small, the microscopic, the hidden world. See in your mind and heart how these parts of Nature are as you are: equal, complementary, soulful.

Then, contemplate the existence beyond any of your knowledge of Nature, a deeper, fuller sphere of Nature and consciousness. If you imagine a connection to what is unknown, unnamed in the universe, can you translate this gift to growing your conscious relationship with what you know, what is connected to us all, what is us? Use your imagination to blur the lines you may have created to separate yourself from that which you perceived as not you, from Nature, from yourself. All is one. You are all. Even what is unknown, and beyond.

iRewild

Nature has given us the power of thought, the ability to learn, distinguish, name, and separate what is before us. Let go of this power, this ability. Imagine there are no names, no categories, no separation in our minds and hearts. The power of merely naming is far beyond what it seems. Remove the names, the separateness. What is, is. That is all. Just existence. Wherever you walk, or whatever you think about, is unnamed, no different than you and what you experience. If this is all you know, might all there is be aspects of you, not beyond, but within you, within all of us, connected to all, known and unknown, visible and invisible, near and far?

Reflection

Week 9

Imagine

Spring thrives with life and mystery as misty green landscapes disclose new, exciting possibilities. Summer is a time of bursting vitality, its fields painted with vibrant flowers, and ancient trees stretching for the heavens. Fall is a breathtaking kaleidoscope of colors, while winter is silent with slumber and tranquility. Each season reveals Nature's creativity, intelligence, and beauty. Each season evokes earthly experiences that live beyond words and echo within us for eternity—transforming us.

Imagine Nature as a living organism, constructed by invisible threads and a unifying consciousness that is the source of all life, the intellect inside Nature. Imagine unseen threads crisscrossing the Earth, forming an interconnected matrix that holds our world together, that holds all beings within it. A matrix containing every single thing's life force, together with the intention, the fear, pain, or joy, of all thoughts existing from the beginning of time. Imagine thoughts as energy signals. Thoughts which formed our world. Thoughts which underpin our existence. Imagine Earth embodying a cosmic consciousness, masterfully recognizing within her intrinsic nature the very essence of the universe.

Thoughts are things as real as dandelion flowers, sky-high redwood trees, river rocks, and snow geese. They are even more real than the material world because thoughts have the power to create and shape reality. Imagine humanity lying in the borderland between the microcosm, the subatomic waves and particles, and the macrocosm, to the furthest edges of the cosmos. We live our lives in an interactive world where the intentions of our thoughts are embedded in the substance of our world.

Like a drop of dye in the ocean travels outward into the Atlantic, everything we do in life, every movement, every word, every thought, has the energy to help, heal, or hurt. The energies that we discharge in the microcosm make an even bigger impact on the macrocosm, the greater order of our universe. Imagine your life energies as agents of creation, working within an absolute unity that becomes the timeless, life-giving essence of our universe. We are one and the same, each found in the other, inseparable, all creating our reality.

Question

Scientific studies show that our thoughts and outlook on life directly impact our health. Every thought triggers the release of chemicals in our brains that have the potential to affect us both physically and psychologically. Negative thoughts can release chemicals that are harmful to your health, while positive thoughts release chemicals that favorably influence your mind and body. How can you maximize your opportunities for positive thoughts? How can you change your attitude to see and respond to situations from a more positive point of view?

Explore

Earth's natural world is an act of co-creation. This is why Nature mesmerizes us, inspires us, and invites us into relationship with her. Stories are born from these acts of co-creation between you and the world surrounding you. We are each a living, breathing, active part of these stories. Stories that help to deeply root us to a place and see the world as alive. These stories meaningfully connect the ancient past to our present. Just you living where you do introduces new possibilities, contributing to or altering Earth's story.

Create or obtain a map of your surrounding landscape, where you live or frequently go on outings. What memories do your surroundings hold? Write your answers, or make notes to your map, for the following questions:

Are there preexisting stories about your area? What are they? Who were the land's ancestors?

What events have taken place? Are there any landmarks? What are their names?

What animals take part in the stories? How do the animals shape the stories?

What animals live in the area now? How have these animals changed the landscape?

How did past glaciers and volcanoes co-create the landscape?

What stories from your area claim your attention? What stories seize your imagination? What about your landscape seeps into your dreams?

What archetypal figures or recurring images do you see expressed in the land's characters and stories? How have these archetypal figures influenced you—your sense of purpose, meaning, motivations, creative potential, and desires for originality?

Research the etymology of your name. Does the meaning of your name parallel any of the stories of your landscape?

Do you see any themes weaved in the stories across time? Do you see any of the stories' threads pulled into and entwined in your world today?

The Earth's stories and memories are alive within the land where you live. Just as you walk within the soul of the world, the soul of the world also exists within you. Just as you move about your landscapes, the land's essence flows within you. What will you do knowing that you are written into a chapter of the world's story for this land? What kind of character do you want to be, knowing that you will remain in Earth's memory, forever?

iRewild

We are conditioned by our families, cultures, and society to believe only certain aspects of reality. Yet, scientists tell us that, on the subatomic level, merely observing a thing may change the thing being observed. Why do you think people tend to believe that thoughts are less real and important than words, and words are less real and important than behavior? How do you see your consciousness relating to the universe, to matter and energy? How do your actions, thoughts, and words affect the macrocosm?

Reflection

Week 10

Imagine

In the mist of the early spring morning, dew covers the ground, softening your steps as you walk slowly beneath the tall forest trees. On the edge of the trickling stream, a mother red fox silently observes her four young pups rolling, pouncing, and chasing one another in circles around the budding maple trees and large, moss-covered rocks. Her keen senses are attuned to her environment. She notices all the hidden aspects that might otherwise be neglected. Her level of awareness invokes your curiosity and wonder. You stand a while and sink down into your thoughts. You reflect that living in an urban area has isolated you from connecting with Nature. You consider how the countless struggles and responsibilities of life have slowly compelled you to withdraw your attention from sensorial experience. Perhaps these two things have gradually numbed you to feelings of appreciation, empathy, and love, maybe even causing you to feel indifferent toward others, and the planet.

Question

Where in you is this numbness? What does it look like, feel like? If you could give it a voice, what would it say? How has your numbness kept you from experiencing life to the fullest?

Explore

Find a spot to sit in a natural setting or near an open window. Choose something in Nature that you can physically see, hear, or feel. Remember that air, sunlight, rain, and clouds are available to you if you can't be outside near a plant, tree, river, or animal. For a few moments, close an imaginary door to your reasoning and critical-thinking mind. Tell these potentially overwhelming thoughts that you will listen to their perspectives after the exercise.

Set an intention to be present in the moment. From your heart center, imagine a bright beam of light linking your heart to what you have chosen. Then, imagine that your entire being is your soul. See your soul beaming like a beacon of radiant light, connecting with your chosen entity in Nature. Ask yourself what qualities, patterns, and rhythms you would have if you were this being. Embody each quality you observe. Notice any changes in how you feel, think, or respond to what you are linked to. Take a moment to feel like you are in relation with this "Other." For example: the energy and vitality of a leaf absorbing warm sunlight; the calmness and generosity of a drop of rain; or the freedom and lightness of the wind. Move like they do. Feel their presence in the world. Feel your presence and the intimate interconnection and relationship with this Nature-connected moment. Feel a deep sense of gratitude for every phenomenon that you notice, for you are experiencing the sacred in Nature, rather than a commonplace event that is often taken for granted. Thank Nature for allowing you to share in the world as you restore your sensitivity to all life.

Repeat this exercise often and be intentional about noticing the natural world that surrounds you throughout the day. These day-to-day moments shift Mother Earth's story, expanding our view of, and connection with, the natural world. Learning to acknowledge an-Other is empowering and revitalizing. You will find yourself perceiving things that have long gone unnoticed.

iRewild

The fox's keen senses navigate illusions and provide a much more expanded and powerful way to perceive the world. Deep attention and listening ignite curiosity and understanding, which allow us to see through the eyes of an-Other, through the wordless murmurs of the natural world, and then to connect completely and meaningfully with that experience. We pay attention to all the dimensions of an-Other. We listen for the soft sighs that are too buried for words. We listen to give them a voice, a chance to be seen, to feel that someone understands their struggle. We remember that silence can appear like all is right, and that appearances can be deceiving.

Imagine, as you look back to the stream and peer into the foxes' eyes, what are the fox and her pups longing for that they can't put into words?

Reflection

Week 11

Imagine

In a picture of a young boy sitting next to a tree, one of the tree's branches is an arm, which encircles the boy. Under the image, the boy says to the tree, "The world is so loud and makes so many demands. Sitting next to you doing absolutely nothing means absolutely everything to me."[3]

Question

Thinking about this image and the accompanying words, ask yourself how Nature extends a loving hand to you. How can Nature reveal her heart to you? Is there anything you need to do to develop this deeper relationship with the natural world?

Explore

Take a deep breath and place your palm over your heart. Move into your heart space. Breathe deeply in and out of your heart for a few moments, then visualize yourself in your favorite place in Nature. Continue to breathe slowly and rhythmically, connecting with Nature. Continue in this space for as long as you feel comfortable. Make notes on your experience.

iRewild

Can you feel Nature's embrace? How does it make you feel? How might you make this deeper connection to Nature more alive in the world?

---

3   Francois Lange, SketchesinStillness.com, May 23, 2021, https://sketchesinstillness.tumblr.com/post/652013460069171201/the-world-is-so-loud-and-makes-so-many-demands. Used with permission.

Reflection

_____

_____

_____

_____

Week 12

Imagine
The night sky fills us with awe and wonder. When you step outside into the night air and look up through the darkness, you see an ocean of black accented by infinite stars, and a moon that is calling to you. Whether it's the Snow Moon or Blue Moon, crescent, quarter, or full moon, from sunset to sunrise, the moon's light bathes you with her silent, bluish-white stream of purity and innocence. Known as the Goddess of the Heavens, the inspirer of human love, the holder of hopes and dreams, the pathway to wisdom and understanding, the moon is a wandering, luminous sphere symbolizing the phases and evolutions of ourselves. She reminds us that everything in our world has a cycle, and darkness does not extinguish the beauty of life. At one time, the shepherds and seafarers looked to the moon and stars to find their way. But tonight, together with Earth's oceans, the moon is pulling on your heart, her radiance reaching down into the hidden depths of your being, moving with you in an intimate dance to bring light into your life.

Question

The moon's light takes you on a path of self-unfoldment, the blossoming of your inner world to yourself. We must look deeply within ourselves for transformation to take place. As the moon looks at you, how would this radiant being describe the true you, the person on the inside that the moon is illuminating? Not the persona you present in public, but the inner you. Draw a moon, and, within that image, try to write the most precise, honest description of yourself possible. Like the moon who is continually changing, what do you want to transform within yourself?

Explore

Find a comfortable spot to sit where you have an open view of the moon. If it's cool outside, bring a blanket or dress warmly. If you can't get outdoors, find a window with a clear view of the moon and turn off all the lights so the moonlight can illuminate your home. Close your eyes, tilt your head toward the moon, and take three deep breaths. Visualize your body not being something solid, but being made of living, intelligent, radiant light energy. See the moonlight bathing, embracing, connecting with your light energy. Visualize your light body getting brighter and brighter, lighting and washing away the darkest corners of your life. Feel the moon's energy restore your energy. Stay with this image for a few moments. Open your eyes, relax, and gaze at the moon. Simply look at Earth's beloved partner for several minutes. Immerse yourself in her light. Open yourself up to receiving her light. Welcome her energy. Unite with her. Feel her purity, clarity, and freshness join with your beingness to revitalize your mind and body. Every time you see the moon, honor her, and remember this experience.

iRewild

Myths and legends talk about ancient civilizations associated with the moon's unique meanings and powers. For a moment, imagine yourself as a part of an ancient tradition. You are the protector of plants, animals, and waters who dwell in the natural world. In Nature, all beings work together, each have their special job that helps to sustain our planet. What beings are you drawn to? Why? What role do you fill that helps to keep the natural world healthy?

Reflection

Week 13

Imagine

What if you opened your heart to the possibility that you are embedded in an unfolding, dynamic relation between everything and everyone that you meet—plant, stone, and two-legged beings? What if you opened your heart to the idea that all is connected at the deepest level—rock, tree, and winged beings? What if you opened your heart to feeling, hearing, tasting, and touching this deeply special space where you and the world meet, where you can experience the communion of rose, boulder, and crawling beings? What would it be like to know that you can enter a vibrant space of co-creation with any being that comes into relationship with you— four-legged beings, pebbles, and the vegetable kingdom? How would it change your world to know that this is a space where there is no hierarchy, no expectation, no striving for solutions or answers, that this is simply a space of mutual appreciation, recognition, and love? What would it be like to know that the world reaches out to you in every living moment, asking you to sing and dance with it in all of its glorious complexity, multiplicity, and vibrant color, that you breathe life into each other? Can you hear what swimming beings, ferns, and fossils want to say to you? Are you able to hold and honor what the world is offering you, how the world is communicating with you?

Question

Just as you see and hear the world and its beautiful subjects, allow yourself to contemplate that the world sees and hears you. What would it be likvve to experience the world through another being's senses, to understand the world through its perspective, to appreciate another being for being exactly who they are? Take a deep breath, sink into your heart, and immerse yourself in the life of a caterpillar, for even the caterpillar sings:

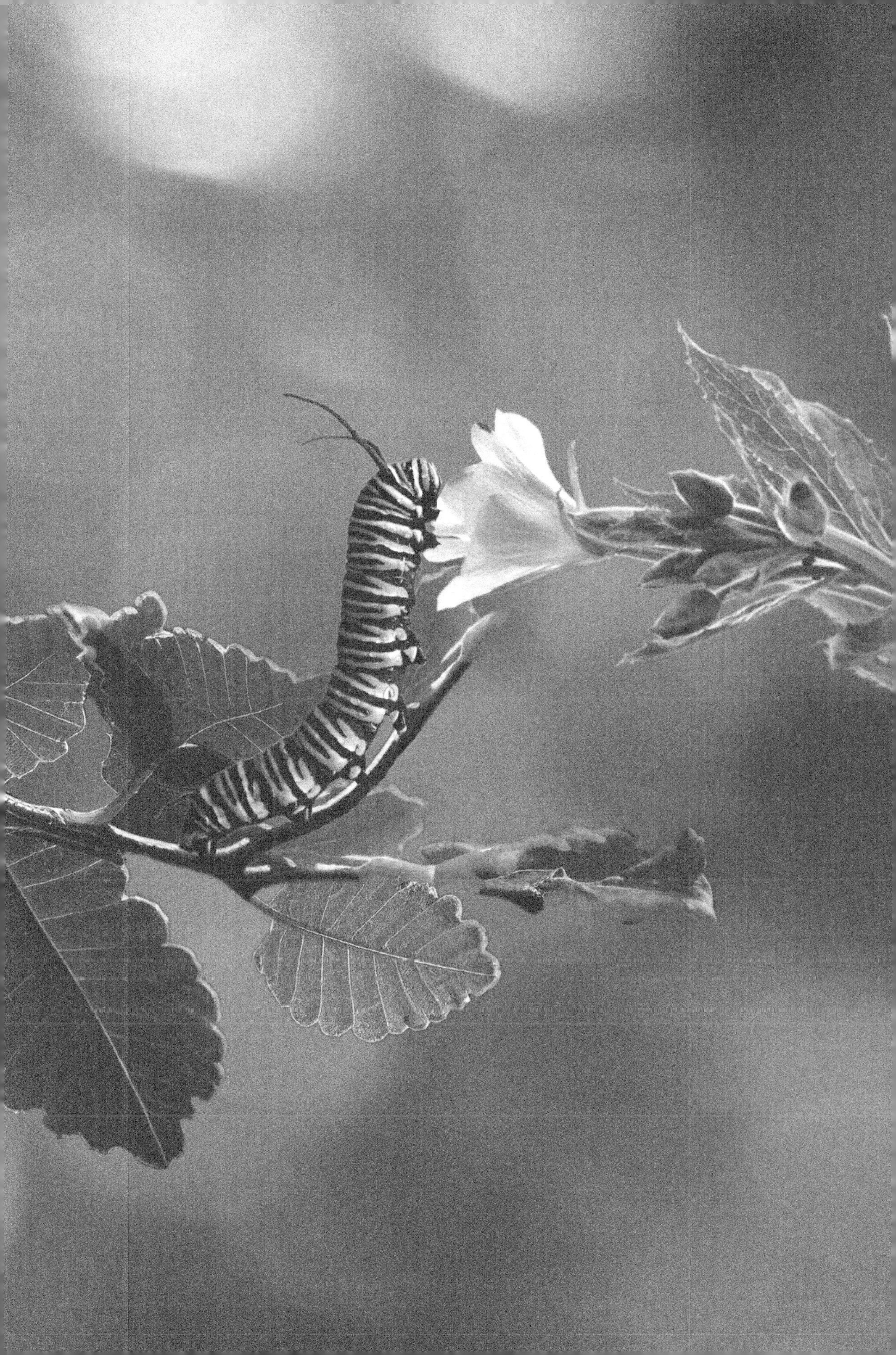

I am Caterpillar.

My home is the broccoli plant in the vegetable patch.

I live here with my family; we are happy.

In times of plenty we feed until we are full, as we don't know when we will be able to eat again.

I enjoy warm sun on my body on a late summer day;

I shelter from rain drops;

I get scared when dark shadows fall.

Will I be eaten today, or will I live to eat some more?

The broccoli plant is my home;

The vegetable patch is my village;

The garden is my country.

I haven't travelled very far,

But I love where I live.

What would it be like to put yourself in the world's shoes, to see yourself as the world sees you? How would this change your relationship with the world, your actions in the world?

Explore

Move out into Nature. Find a quiet spot where you will not be disturbed—perhaps under a shady tree, or next to a babbling brook. Sit down, make yourself comfortable, and give thanks to this place for accepting your presence, for seeing you and holding you. Take a deep breath and invite Nature to speak to you. What do you hear, feel, see, smell, taste? Hold the possibility that this is Mother Earth communicating with you, breathing life into your whole self. In this sense, what do Mother Earth and her offspring want you to know?

iRewild

You and Nature are so utterly precious. What would it be like to consider that when Nature sees you, hears you, holds you, your consciousness is brought into being, that you are brought into being by Nature? In the spirit of reciprocity, how might it change your perspective to know that you are responsible for bringing the story and beauty of rose into being, for bringing the fragility and vibrancy of caterpillar to life, and for bringing the deep wisdom and enduring presence of stone into being? Sing a song, write a poem or a story, or draw a picture that honors Mother Earth, that lets Nature know that you see her, breathing life back into her. For without your love, appreciation, and gratitude for these beings, without you seeing and hearing these beings, could they ultimately disappear?

Reflection

Week 14

Imagine

In the hours of early morning, you sit near the top of the mountain and watch clouds touch the ground below. Piercing through the mist like islands in a sea of gently rolling clouds, the surrounding mountains appear to be distant old friends, transforming the landscape below from merely beautiful to breathtaking. The ferns, trees, and plants seem grateful to collect the small water droplets in the fog to relieve their thirst, adding to the other-worldly feeling. From moment to moment, the clouds lift and melt to reveal peaks and valleys marbled with streaks of golden sunshine, transforming the wondrous terrain beneath into a world caught between reality and illusion.

Question

As you gaze at the rolling clouds and mist in the valley below, you wonder, in this era of technology, how you have gradually, often unconsciously, stepped out of the natural world and into a virtual world. How has technology become your lens to the natural world? How has technology impacted your relationship with Nature? Is the natural world surrounding you really how you see it, or is that just how the modern world presents it to you?

Explore

Find a quiet place to sit in a natural area without any technology—your back yard, city park, or forest. Close your eyes for a moment. Take a deep breath and open your eyes. Look around at your surroundings and observe what is happening to the landscape around you. How much, and what, specifically, is due to seasonal changes? Are the flowers blooming or going to seed? Or are they dormant under a winter blanket of leaves? What is happening with the birds? Are they in flight or perched in nearby trees? Are they feeding their young, singing a song, or avoiding your presence? How are the squirrels behaving? Are they running away from you, playing along the tree branches, or feeding on the ground? Are you standing on a plant, grass, or moss, or leaving footprints in the soil? Is the ground moist or cracked and dry? How is your presence impacting the trees, animals, plant life, or insects that live here?

iRewild

What does your heart tell you about the natural world that lives through you?

Reflection

Week 15

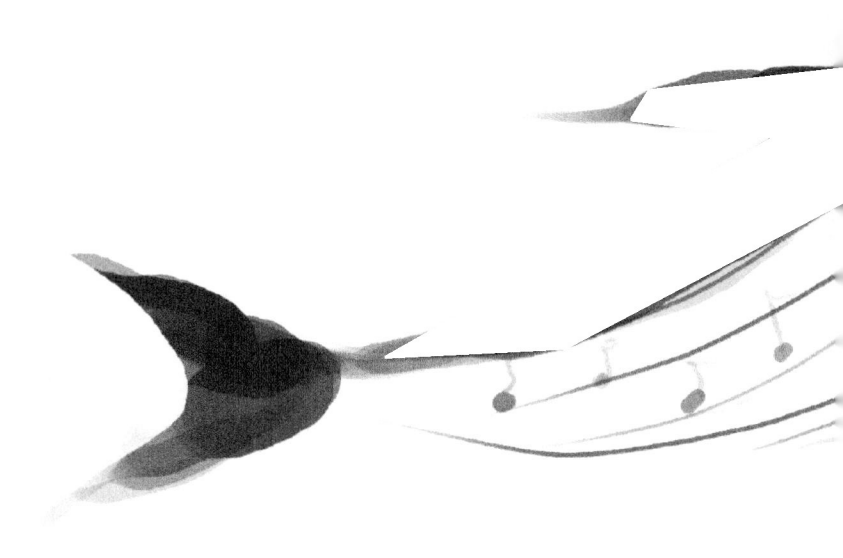

Imagine

Blue whales are the ocean's largest creatures. Like us, they sometimes travel in groups and, other times, alone. They communicate with one another in loud, low-pitched moans, or mating songs, that are easily audible over long distances.

Many whale songs are long, complex, and melodic. They compose fragments into pieces of music that are intricate and evolving. Other whales in the group learn the phrases and sing along. Over time, like revered jazz musicians who master the art of improvisation, fellow whales collaborate and become co-creative forces, adding fresh melodies that eventually become new songs. Listen closely and you will hear the mysterious and eerie serenade of the sea resonate through the ocean waters, slowly fading just before the whale surfaces.

Question

In their pods, whales see one another as sacred beings, dwelling in a realm of divine goals. Is each breath and step you take a dance of grace and a sacred journey?

Explore

Breathe the world into your heart space. As you breathe out, breathe yourself back into the world. Feel this participatory relationship growing in your heart space. In this space, bring into your heart's awareness the living, breathing ocean. Imagine that you are swimming in the ocean's deepest depths. As you continue to breathe in and out of your heart, you see a pod of blue whales ahead. You move closer, in awe of their immense size, their incredible grace, and deep resonant song. You reach out your heart to them and connect. Continue this relational breathing for as long as feels comfortable. What particularly claims your attention as you float in this space with these magnificent creatures?

iRewild

What has it been like to gift yourself the time of reaching out in loving participation to one of Mother Earth's creatures? How do you feel that this contemplative action has affected your life? Moving forward in your everyday life, can you gift yourself the time to be in contemplative participation with more of Mother Earth's beloved creatures? How could this practice change your life and the wider world?

Reflection

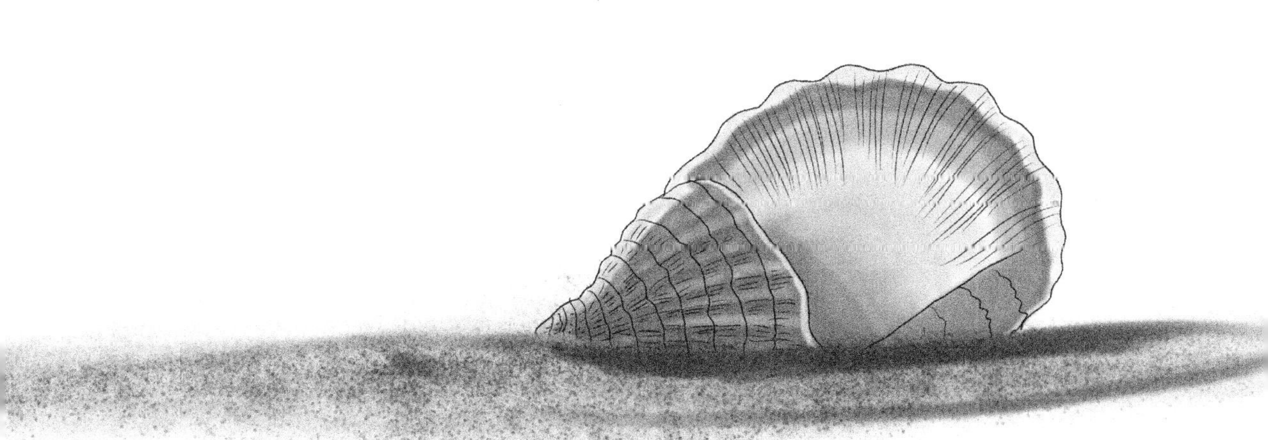

Imagine

When you look up into the night sky, every star you see is nestled in the Milky Way, the galaxy you call home. Winding like a silver river across the dark sky, it is so intensely studded with glimmering stars that it is possible to see the spiraling clouds of dust and gas. The Milky Way seems to be a borderland between one reality and the next—a gateway letting in a shower of stars. It is as if the infinite universe reveals itself under your inquisitive gaze, inviting you into the infinite universe within.

Question

Imagination has no limits. It reveals the deep longings of the soul. You yearn to get as close as possible to the mystery. When was the last time you used your imagination to envision the different aspects and dimensions of reality? We share our essence with the wider natural world. Our sense of our soul is interwoven within the larger soul that encompasses all of Nature. Using your imagination, what is soul to you?

Week 16

Explore

For this exploration, you will need paper, pen, and colored pencils, crayons, watercolors, or any medium of your choice. Then find a place where you can include Nature. Step outside and, if possible, lean against a tree or sit by a stream.

Let your soul guide you into your imagination. In this space, let go of logic-based thinking and return to dream images as vision-based knowing. Focus your attention on your soul. See your soul as a living being through which you exist. You are an extraordinary expression of your soul. Imagine your soul's shape, color, how it feels—speak to it, experience it, strive to understand it. Where and how does it live? Visualize how it is connected to the natural world surrounding you. Try to see through the eyes of your soul. See your soul both as part of the whole and as permeated by the soul of the world.

Write and draw what you see as an image. Do not change or omit anything from what you see, for that would be telling only part of your story. You could be turning your back on the most important elements for understanding your soul.

When finished, honor your drawing. Look at the image with a new perspective, beyond the surface. Enter into it. What does the image evoke in you? Fold it up and put it in your pocket. Carry it with you. Every now and then, take another look at the image. Do your movements, your thinking, your feelings correspond with your drawing?

iRewild
In what ways does your soul long to restore its relationship with the mysteries that govern the universe, the underlying forces of Nature, and the soul of the world?

Reflection

Week 17

Imagine

When you were a child, perhaps you ate mud, played with worms and sticks, ran barefoot in the grass, and used your imagination to create dens and safe spaces in Nature. Were you disgusted by, or afraid of, insects? Or were you curious and eager to learn and understand the world around you? Did you splash in puddles, jump on mossy mounds, run with butterflies and fireflies, and talk to the daisies, roses, and red robin? Did you give names to the plants around you and refer to animals as friends? Perhaps you ate blackberries and cherries from the wild without bothering to wash them.

Question

When did you lose your innocence and connection with Nature? How do you know that you have lost them?

Explore

Think about a natural place you played in as a child. Visit this area, or somewhere similar, and re-enact your natural play from all those years ago. Explore the habitats with your curious inner-child. Pick up mud, put your bare feet in a stream, hug a tree, make a den out of sticks and leaves. Introduce yourself to trees and squirrels and maybe, just maybe, they'll tell you their names. Follow an insect and see where he goes and what he does. Spin around with the dandelion and sycamore seeds. Climb some rocks and shout like a wildlife warrior to announce your connection to Nature. Use your senses like a child; smell, touch, and listen with uninhibited curiosity. Go with the flow and follow the desires of your Nature-connected inner-child.

iRewild

If your playful inner-child could talk to you through the rattling leaves on the trees one breezy day, what would they ask you to bring back into your life to regain Nature-connection?

Reflection

Week 18

Imagine

As a living being, participating in the world's unfolding, you are always in communication with the organism that is you, and the vast organic entity of which you are a part—the Earth. In this context, communication is an exchange far more primordial, more constant, than verbal exchanges between human beings. This exchange is a living, reciprocal interaction between two living presences—between your living, breathing being, and the vast living, breathing Earth. In this sense, perhaps this exchange is not just communication, but communion—a sort of sensuous immersion in dynamic relationship—communication beyond words.

Question

In your day-to-day life, what do you understand by communication? What does the term "relationship" mean to you? What are the dynamics in both? Moving beyond your day-to-day reality, are you able to hold the possibility that you can communicate with the wider world? How do your understandings of communication and relationship change in this context? What might you need to do to expand your awareness to hold these ideas?

Explore

Close your eyes and breathe slowly, in and out, through your heart. If you have difficulty with this, bring something or someone into your awareness that you have a loving relationship with. Sense the energetic exchange as you feel the love between you and the being that you've brought into your awareness. Then visualize yourself in your favorite place in Nature. Continue to breathe slowly and rhythmically, connecting in relationship with the natural world within which you are held. Continue in this space for as long as you feel comfortable. Make notes on your experience and think about how you could move this knowledge and understanding into a wider conception of your growing relationship with the natural world.

iRewild

As you are developing your relationship with the natural world, if Nature could give you a message, what would she say to you?

Reflection

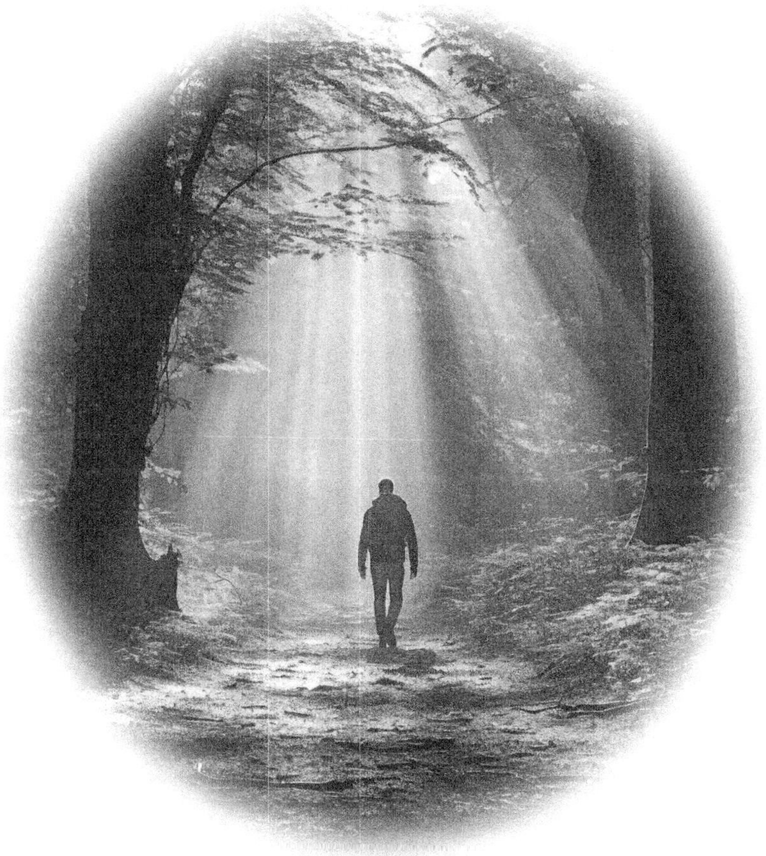

Week 19

Imagine

The hoverfly, a seemingly furious, yet gentle, beauty, is fascinatingly fast and has an armor of colors to mimic her close allies, the wasps and bees. If you're chosen, one may land on your hand, or hover in front of your eyes, as though she wants you to connect with her tiny, adventurous soul. If you look deeply into the glassy eyes of a hoverfly, you may be lucky enough to see the sky twinkling back at you.

Do animals sense kindness? Or is it a deep curiosity because you glow with compassion and send out a signal, a vibration, of openness and oneness? When you imagine yourself from the perspective of an-Other looking back at you, whether through the eyes of a lizard, fish, rabbit, or plant, you can then begin to open the door to Nature-connected empathy. If you can see yourself from their experiences of you, and sense the emotions and feelings they have toward you, in your presence, then you can begin to understand their view of you, and why some people are chosen to land upon, and some are not.

Question

Your imagination is much more powerful than simply a tool to create stories. The imagination is a gateway to communication without words, to experiences with an-Other. Try it yourself—how would the hoverfly wish to be seen by you? If she landed upon your hand, flying curiously up to meet your gaze, what might she be trying to tell you? When the hoverfly looks at you, with kindness, curiosity, and all the wisdom in the world, how would you be seen? Why would she choose to land on you? Would you allow this? Would you let the hoverfly impart a message to you? There's no better time to start opening your heart and sensing with an-Other, Nature, than right now.

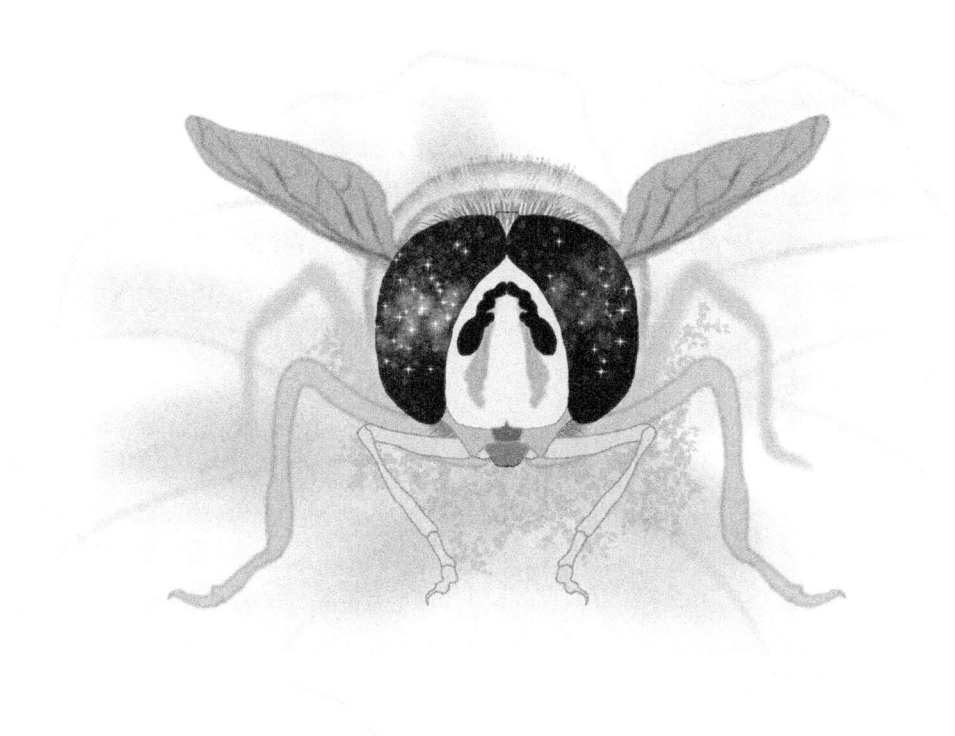

Explore

Have you ever seen flies dance? Like gnats in the last rays of a summer sun, dancing gracefully above a flowering shrub, or fireflies air-skating over a moonlit lake, or bluebottles circling up and down the yard like a rollercoaster? The beauty they portray, and sense of freedom and unity, in synchronous precision with one another, at such speed, is awe-inspiring, breath-taking, and almost unbelievable. The flies invite you to watch them like you've never done before. Sit, or stand, in the stadium of Nature. Admire a world that once, in misunderstood irritation, you may have brushed away with the swipe of your hand. The show is awaiting your arrival, and the ticket is free with an open pass. If the season does not permit the performance, then look for a cast of birds instead. Move your head, sway your body, make a buzzing sound, and copy the movement of their wings. However you choose to embody these moments, feel what it would be like to be with that group of gentle flying beings, as if you, too, were a small flying creature dancing along. Over the coming week, imagine and act out other flying creature's wing movements, like the butterfly, flamingo, hummingbird, or bee.

iRewild

Without words, it may feel nearly impossible to communicate with a hoverfly. Yet, with exploration and seeking to understand, we see that Nature, from the tiniest fungi and amphibian to the biggest baobab tree and mammal that walks the Earth, has unique abilities and diverse methods of communication. For Nature, too, has the need to socialize and share her needs, desires, and eternal wisdom with all that is. How could you commune with a flying insect, like the hoverfly, so they may understand you a little more than they did before?

Reflection

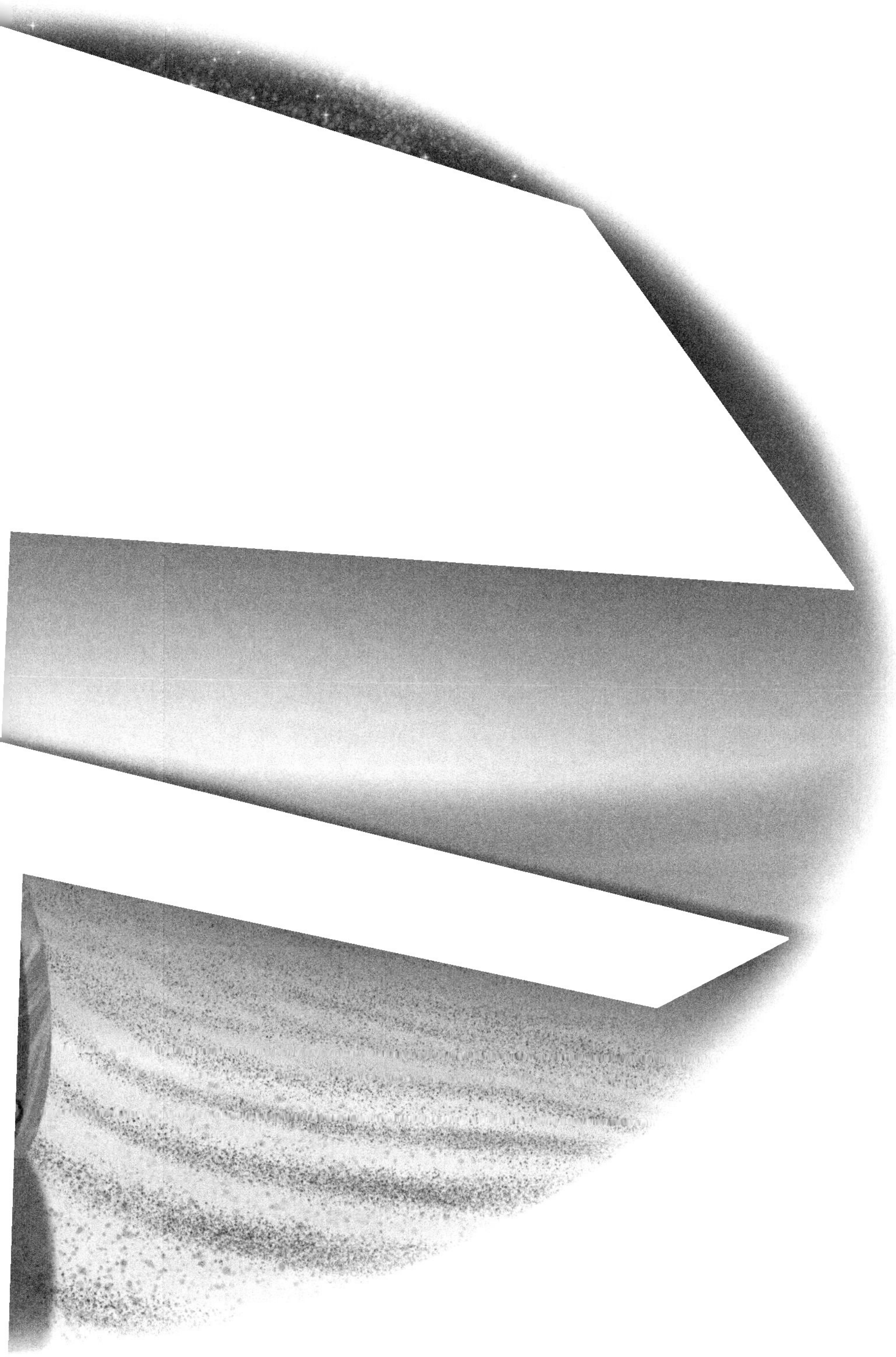

Week 20

Imagine

You breathe in. You breathe out. You are. You were. You will be. Time flows in and around you. Time beckons you into its very being. What is upstream becomes downstream. From the moment of your existence, you create a past from the future as it passes your now. Some say it is an illusion, and all that is ahead and all that is behind is all of each, always. No distinction. No direction. No flow. No march. Just everything. Only our mind's creation gives it a name and feel. Others say it is real, and separate from us. Motion, heat, and change give it existence.

Whether illusion or real, we feel its presence, its pull, its opportunity, its yoke. We can sense it change within us, how we experience it, feel it slow, speed up, stand still, but never stop. It is beautiful in its ability to excite us, provide us anticipation and hope, and yet it is relentless. Never far from us, it can hover, envelop, even caress, but will never leave us, until we leave it. And then it disappears. Forever.

Question

Our connection with time has long been the subject of poets, philosophers, physicists, and storytellers. Our language is time-dependent, sensitive, structured. For something so present, so ubiquitous, we find ourselves wanting when we try to explain it, as if we are squeezing sand or water. How does time affect our feelings? Is it necessary for us to have these experiences? Is the temporal aspect of our existence necessary for us to be human? To be at all?

Explore
Much of our thought relates to our wish to be untethered from time: if only we had more time, less of time's restrictions, the choice to do without it. We have all experienced its seeming fleetingness, stubbornness, refusal to pass. What would it be like to be beyond time? To transcend time? What if we let go of time? Maybe it is our will to hold on to time that gives it the ability to grip us. Some wish, pray, for eternal existence. Others say this would be a worthless experience.

If you were eternal, what experiences could you have? With eternity, time may have no meaning. Would we remain human? Would Nature have any meaning? If you could not experience time, could you experience anything? Time may be necessary to feel joy, sorrow, empathy, connect to an-Other, have a conscious relationship with Nature. The soul itself, thought to be eternal, may not be able to exist if it is not tied to time, and, therefore, to aging, death, and rebirth. Mortality may be necessary for eternal life. The concept of time, the reality of time, the limits of time, may be what ensures our ability to experience consciousness, a conscious relationship with Nature. Forever may lack meaning without time to give it substance, and time may give forever its existence.

Find time to take a walk. Maybe a different kind of walk for you. Remove your watch. Leave your phone behind. And then look and listen and feel. Experience all that is with your senses without reference to time. Reboot your existence. If time does exist, join with it as it is in Nature. Nature's time is our time. Can you join with the cycles of time and let Nature take you beyond our narrow view of time? Not the existence of time, but its weight. See if you can use time, relate to it, differently, and transform yourself by bonding with it, and thereby freeing yourself from it. The closer you are to time, the less you may be bound by it. Take away any power it may have over you and you have no reason to avoid time or run from it.

As you think about how you move through your day, your life, your thoughts, ask yourself how you see time, how you relate to it. Is it your friend? Your foe? Encourage yourself to develop a less quantified relationship with it. It is not numbers. It is not limiting. It is what gives you all that you are. It is what gives you a bond to all other things.

iRewild

If change is necessary for time to exist, either as an illusion or independently of our perceptions, imagine a changeless world where time has no place. Can you sense any beauty? Time may be necessary to understand our existence, our connection to the wider natural world, and to an-Other. If we have been cleaved from our connection to Nature, might Nature's time be the answer to reconnect with our counterparts in Nature? How does a tree experience time? Is it slower? The seasons give it patience. Time pieces have laden us with smaller and smaller increments of time. Have we created time ourselves? Imagine if the tree feels beyond time, free from it.

Many of us dwell on what will happen after we die. It can be difficult to comprehend, to manage. Try something else: imagine where you were before you were born. Were you bound to Nature? A part of Mother Earth?

Reflection

Week 21

Imagine
The ocean has an irresistible way of beckoning us towards her indigo waters. Her hypnotic waves wash away our tears, footsteps, and burdens. Her azure ethereality provides a sense of peace for a weary soul. The pounding of our hearts seems to pace the beat of her waves, which have been heard since the beginning of time. She is a gateway for freedom seekers, like the albatross soaring across her seas. She is the great unexplored vastness of blue, holding mysteries of the deep, containing the secrets of our globe. Unnamed species glow, iridescent, in the absolute darkness of her eerie, bubbly depths, never seen by the grounded eyes of a human being. Life itself may have begun in her womb. The ocean is not our habitat, yet she captivates us and holds us, like our mother's womb. Her healing salt soothes our aching muscles and bones, and her loving water sends tingles through our feet, as we stroll dreamily along her fluid edge.

An ocean sunset is an image many behold in their imaginations, supporting them in difficult times, a scene that can bring a person to tears, overflowing with Nature's beauty. If you offer a pebble to the ocean, as the sun's rays fade into her lucid liquid, at that moment you can become her artist's assistant, splashing colors into the air like a giant invisible paintbrush. Once the pebble drops, regaining its natural shine, you can allow your imagination to join the ripples, entering a vast blue of possibility.

Question
Lie back and use your imagination to float upon the ocean's silky sheet, bobbing safely, smoothly, fluidly. Find that ultramarine place in your mind's eye that makes your soul sing. Are you in a shallow turquoise bay with palm trees swaying in the breeze? A calm, deep,

cobalt ocean beside a wooden boat? Or a silvery moonlit lagoon? Maybe you're floating above pale pebbles rolling in unison beneath you, or on strong coastal waves close to the beach's foam. Step into her motherly lather and join the pulse of the moving ocean, immersing your body into the womb of her water. Let yourself be held by her monumental well of wisdom. What questions would you like to explore with her? After all, she has been here since the beginning of time.

Allow the sea to speak with your soul, to pass waves of inspiration to you. She may even remind you of long-lost dreams, like an old treasure-laden boat sunk in a watery wilderness, waiting to be discovered by the person who bravely crosses a personal ocean in search of their own hidden jewels. Let her water stir up your emotions, like the sand that lies beneath your feet, turned by the ebb and flow of her alluring tide. Smell the sea breeze and ignite your inner senses as you are held in the motherly arms of the ocean.

Explore

If you cannot be at the sea's edge, surf elsewhere for relaxing ocean sounds. Does it resonate with you and call, in rhythm, to your heart? Imagine the warmth of sand or stone on your palms and fingers as you lay back. Listen with full attention and sensory imagination, just like you're there. Your toes touching the warm salty water and your head raised up towards the sun, framed by a heavenly sky. Watch the lapping water meet your gaze. Notice changes in its direction or pace as it flows towards you then back out again to sea in spellbinding repetition.

How does this deep harmony affect your body sense, your breathing and heart's pace? Walk into the soaking cradle of the ocean. Allow your soul to be free. Ride the waves of humanity's great emotional journey, rising with the peaks and falling with the troughs of the waves' pure energy in motion, or relaxing on calm sea days. Let her wash away your troubles, reborn as you emerge with water trickling down your back, sparkling in the sun, landing on the shells and pebbles below you like shedded scales from a marine iguana's skin.

To keep the ocean's placid qualities close by, no matter where you are in the world, notice the representation of a calm sea in the morning sky, stormy waves in grey clouds, deep sea blueness in a lake lined by teal blue flowers. Let the blue remind you of the whisper of a sapphire sea.

iRewild

When you put a seashell to your ear, you're surprised and mesmerized by the sound of the ocean, like the sea's voice is mysteriously captured in a pearly spiral shell that once crawled along her depths. Like the shell singing the siren's song of the sea, what else may you not fully understand in the universe? Face towards the direction of the ocean—maybe she is somewhere on the horizon or right in front of you—and contemplate the following:

Philosopher William James suggests that people are like islands, appearing separate above the water, on the surface, but linked together deep below, connected in the depths, at their very foundation on the ocean's floor. Though our individuality builds "accidental fences," separating us from each other, and our world, our psyches are interwoven in a "continuum of cosmic consciousness," "a mother-sea or reservoir."[4]

Explore in your personal vastness what this quote inspires and raises within you and your deep connection with an-Other.

Reflection

---

[4] William James, "Confidences of a 'Psychical Researcher,'" The American Magazine 68 (1909): 589, https://babel.hathitrust.org/cgi/pt?id=inu.32000000494403&view=1up&seq=603.

Week 22

Imagine

Through the layers of the Earth, we see Nature's vast family, our family, in time. All of our animal and plant cousins, upon which we continue the story of life. Relatives in kind, and of a different world. Like a family photo album, we can reach ever further away from ourselves, and yet closer to our nature. As the pages turn, the memories fade, then disappear into questions of who and where and what. And then into an-Other, fascination, and the unknown, unrecognized, unseen. Our extended family of cousins throughout Nature appear to be not of us, maybe no longer of anything. Page after page, turn after turn, we see less of ourselves, but more of who we are, and what we are made of. Our family grows into an enveloping fog of connectedness upon which we rise in harmonic voice.

Question

Our family history is deeper than the photos we have of our parents, children, and friends. It lies beneath us in the soil and rock, stratified, compacted. If it was as deep and distant as the time accompanying it, we would only have our imagination to even contemplate it. But Nature has gifted us a path to touch our cousins, feel them, bring them to our present family. This is the result of so much family history and time, each building on that which preceded it, pushing it together, so that we may enliven it in our time. We need to dig only a little to connect with our past. Are you a part of the worlds we discover deep in our shared histories? Can you imagine your existence without this extended family? Would it even be possible?

Explore

Find a peaceful place to lie down and look up for a moment. Then close your eyes and allow yourself to melt and seep into the soil, joining with Mother Earth, slowly, continually, down through the layers, meeting your cousins in soul, their lives, their times. And then leave those you have connected to, as you journey further down, and join with more distant cousins, deeper and deeper. Imagine you are as you are now, in your present form, and then imagine you are of the form of the cousins you bond with as you pass deeper through your photo album provided by Nature.

Who is your family? Can you sense Nature as building your family, strengthening your bond? Do you feel gratitude for Nature conceiving all that has been, is, your vast family?

After a time, open your eyes and look to the sky. Journey through the space above you and sense the layers of the universe that created the Earth, and your family, your cousins. Maybe the deep history of family you have discovered, experienced, is but the surface of your connection to a deeper family history, still unknown, not yet conceived. Imagine the universe cradling you as its creation, together with all that surrounds you.

iRewild

We have journeyed through the layers of Mother Earth and the layers of space and found our deeper family, extended beyond our form, our thoughts, and reaching back through the ages and into our imaginations, into other worlds. Our connections are broad and deep, encompassing all that Nature gives us so that we are not alone. Our relationship to Nature's family, our family, the soul of the world, the universe, allows us to contemplate that there is but one family, our family, all of Nature's family. The universe reveals to us this connection, this unity. There is thought that says we are all one, of each other, for each other, connected, of one consciousness, with each of us being an expression of this consciousness. If this is so, then bonding with our family, as we now know it to be after our journey, is a learning of ourselves, the innumerable other aspects of ourselves, through the layers of Mother Earth, the layers of the universe, and as part of the consciousness of everything. One self, one family, always. If you are an aspect of all there is, and all there is is an aspect of you, does the word "reciprocity" take on a fuller meaning for you? How will your path through life be affected?

Reflection

Week 23

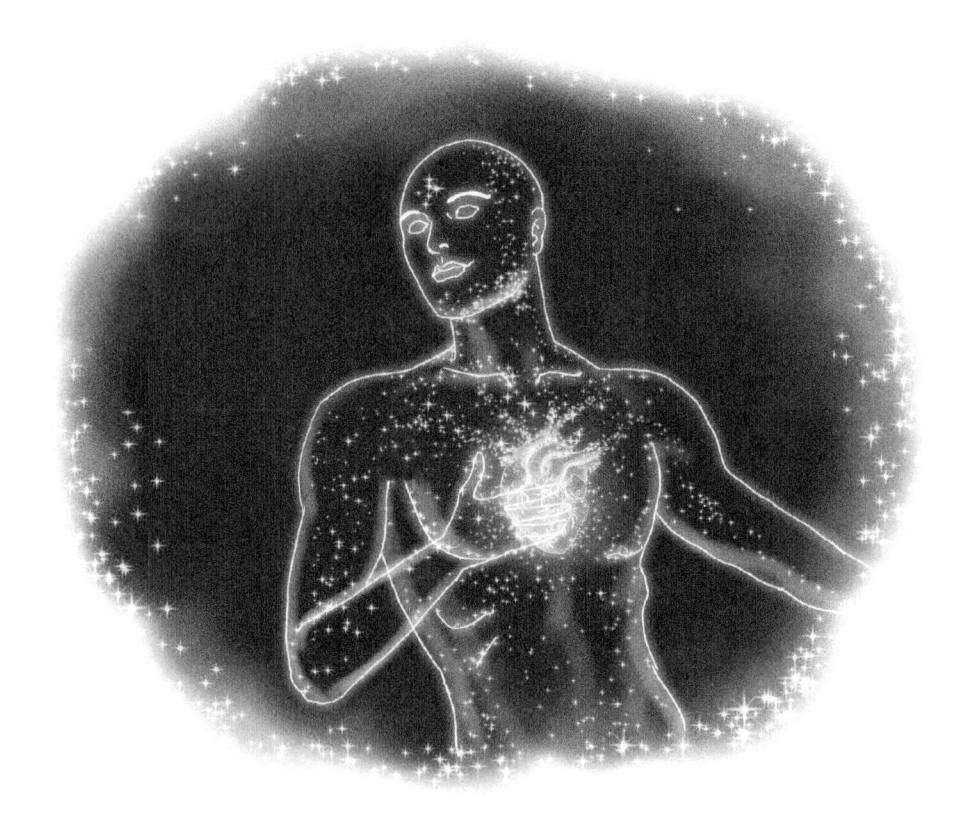

Imagine

Imagine you are going on a journey through the heart's story, the changing faces of the heart through time. The heart has been known in many different ways across millennia, from an organ of intelligence, to the site of emotions, to the seat of the soul, to a figure of sentiment, to the medicalized biological organ of contemporary times. It could be said that the heart has inhabited many worlds and continues to do so, making meaning in the lives of the people within which it has lived across many thousands of years. This is a heart that needs you, in relationship with it, to bring it into being. Can you imagine how these different hearts might have lived within the people of that time, and how each of these different hearts might have shaped their perspectives on life itself?

In our ancient past, and even today, some say the heart is a helical shape, functioning in a twisting, spiraling motion. This is yet another way of seeing the heart, and engaging with the heart, which could ultimately have implications for our own lives. Just as our perception of the heart has transformed in numerous ways over the centuries, bringing forward particular ways of seeing and making meaning in the world, what could be the deeper implications of the spiraling heart that is being birthed once again in our modern consciousness? What might the heart have been trying to teach us throughout history about the ways that we engage in the world? How does your heart live within you?

Question

How do you experience the world? Can you locate in your body from where you experience the world? Take some time to sit with these questions. If you can, move this question beyond your own body, out into the wider world and her subjects. Can you imagine how a plant experiences the world—from where within its body, its leaves, branches, and roots? How a bird experiences the world—from where within its body, its feathers, its heart? How a rock experiences the world—from where within its body, its crystalline structure? While it might be impossible for us to ever know empirically—from a rational, left-brain perspective—your perceptive imagination, which many scholars believe is enlivened through the heart, can take you deep into the world to contemplate the idea that everyone and everything experiences the world differently.

What if you stayed with the notion that everything experiences life in a way that is real and true for them? Might the spiraling heart be teaching us that just because someone's way of seeing the world might not fit into a particular narrative or way of seeing, that does not mean it is not real? Could the heart's way of knowing respect this way of seeing, or open a space for this way of seeing, and consequently be willing to move into messy, uncertain, difficult spaces and dialogues? Can you feel, from the perspective of your heart, that everyone, and everything, matters, experiencing the world in their own ways? In this sense, can you move from "us OR Nature," to "us AND Nature"; from "I'm right, you're wrong," to "this is my view AND this is your view"? Could a philosophical understanding of a spiraling heart expand and deepen your experience of life and offer you possibilities for creativity, different perspectives and unique ideas, to enter your consciousness that may never have had the opportunity to arise previously?

Explore

Slow down. Move your awareness into your own heart, into your imagination. Make yourself comfortable in the space you find yourself in. If it feels comfortable to do so, close your eyes and place your palms over your heart. Now breathe slowly, rhythmically. Breathe the world into your heart space. With every in-breath, you receive and take deep into your heart this amazingly complex, mysterious, beautiful world. With every out-breath, you give yourself back to the world in a reciprocal, dynamic relationship. Breathing in . . . and breathing out . . . deep into the heart space and out again.

In conversation with the world, what is arising for you? Can you, with deep curiosity and a joy-filled heart, wander with conscious awareness into a relationship with the beings around you? What do you imagine they're saying? How do you imagine they're experiencing the world? Does your presence affect their experience? Can you enter into a loving, open-hearted relationship with the beings that surround you? Who is in the space with you? What are these beings saying to you? Can you touch, taste, smell them? What do these beings wish you to know as they speak deeply to your heart? How do they feel about your presence in their world?

iRewild

How does the world look to you through the eyes of your own heart? Is it different than your ordinary day-to-day mode of consciousness? If so, how? Write your thoughts down in a journal. What does your heart have to teach you in relationship with the living, breathing world?

Reflection

Week 24

Imagine

While out walking one day, you notice a solitary acorn. He is a ripe, plump acorn, gifted to the Earth from the towering oak that stretches out above your head. Breathing deeply and closing your eyes, you move your attention into your heart and hear the acorn calling your name. "I have something to teach you," whispers the acorn. You open your eyes and know that he wants to be picked up. In the palm of your hand, the acorn looks so small and fragile. "Yes," he acknowledges, "but I am also strong, and I contain a whole oak tree within me." The acorn guides you to plant him in a perfect place with just enough sunlight, and just enough rich earth, and just enough water. As you give thanks for being in relationship with the acorn, from underneath the blanket of the warm earth, he asks you, "How does it feel to know that you have planted me, yet you may not live to sit underneath my canopy?" providing a welcome home for thousands of creatures, long after your time?

Question

What would it feel like to enter a space wherein you viewed yourself in relationship with the world and received it as a gift? What could enable you to enter this space? What prevents you from entering this space?

Explore

From your heart, move gently out into the world, and connect with the trees in your local area. Reach out from your heart into their world. Many of the trees you see were probably planted from seed by people who are no longer living, who cannot enjoy their beauty as you are doing now. Notice what arises for you as you step into the trees' world. What message do the trees have for you? Make notes in a journal or create a piece of artwork to capture something of this experience.

iRewild

Make a conscious pledge to plant a tree. As you plant your tree, contemplate the idea that a tree's life is often much longer than the human life. In this case, you may never be able to sit under the lush canopy of this fully grown tree. How does this make you feel? Can you plant this tree and wish it a long and healthy life, knowing that it will be oxygenating the planet, and providing a welcome home for thousands of creatures, long after your time?

Reflection

Week 25

Imagine
You find yourself walking by a flowing river with crystal clear waters. You hear a soft voice calling out to you . . .

I am river, and I am as dynamic as can be. I flow fiercely through valleys of vast snow-capped mountains, or delicately in a savannah landscape. Sometimes, I dry up entirely or freeze on my journey. I force through giant gorges or trickle gently into a tropical lagoon surrounded by ancient palm trees. I flow with or without rush, past enormous boulders and fallen branches lying in my way. Sometimes I stray and stagnate into a murky swamp, yet I enhance the green glow of ferns and moss around me. Rains will fall, refreshing my stillness and washing away my old friend algae. Then, I reunite with my whole.

I know I am simply to be, to flow. I trust the process and announce my intention with a raging roar in whirling rapids, or I sing a soft serenade as I stream over silky stones. I am selfless, as is my purpose, sharing myself with you. I am one with every being and place on Earth.

I am more than meets the eye. My body is water. I am your body, your life source. I am ocean, rain, and the eerie mist rising on your fields, or a rolling fog meandering through alpine valleys. I am wispy clouds floating in a still sky and the morning dew mirroring the galaxy and stars above. In one way or another, I have touched everything, including you, from the beginning of time. You call me a sacred source of life. I was revered by your ancestors and I am adored by all of Nature. I wonder if you value me as the red deer does.

From above, you may witness my intertwining web of water. Many say this breath-taking image changes their view of life itself. They see beyond me and recognize reflections of Nature, of themselves, in my wandering waters, shaped like leafless trees or human lungs. In a moment of wonder, you know that I am you, and you are me. You cannot be without me; you are of water.

Question

Close your eyes and open your imagination. Think back to a time you were by a river. What did it look, smell, and sound like? What sensations and emotions arose within you? What do you miss about the river when you are away from it? If the water touched your skin, can you remember how it made you feel? Were you hypnotized by the sun's rays twinkling on the surface like crystals, or did you float weightlessly in its embrace? Did you fall asleep to the calming lullaby or stroll along the banks meeting its kin? As you reflect back to your time with the river, remember what happened to you in its presence.

Explore

Throughout time, many humans have lost their deep connection with Nature and her mysteries and messages carried by the flowing river. Work towards being one with the majority of who you physically are—water. Find a natural flowing body of water and place your feet comfortably within its flow. First, as you wiggle your toes, feel your own water pulsing inside of your body, then become one with the river water rushing past your skin, sensing both bodies of water at once. You are mostly liquid, like a river, flowing together, sharing an experience with each other, in deep Nature-connected empathy. Retain this memory by drawing a representation of what you saw, felt, and experienced in the flowing water.

iRewild

What would river need you to do for water to connect more deeply with you? Nature has a special way of guiding us—let her inspire you through river and the internal and external natural flow of water that is you.

Reflection

Week 26

Imagine

Each morning, whether it's warm, foggy, or raining, you take the time to walk past the patch of rose bushes that line the timeworn stone wall that winds down the hill and into the forest. Each day you stop, gently bring one of the dark pink buds close to your face, and inhale deeply. The sweet floral molecules seem to gently flow into your nasal passages on the way to your lungs, where they cross into your blood and are carried to your heart. In that brief moment, the fragrance seems to take with it all of your thoughts, anxieties, and fears. To your surprise, you release them easily into the wind. At once, your heart and mind rest in awe, and Nature's intelligence deeply inspires you, awakening you to an expanded world.

Once again, you breathe in the freshness and innocence of the flowers. You breathe in the peace and stillness of the trees. You feel deeply in touch with the peace, harmony, and beauty of Nature that surrounds you. For the first time in ages, you are truly present in the moment. The rose scent reminds you of your soul, of spiritual joy—a deep joy arising from the depths of your beingness.

Question

The sweetest things in life are free. Appreciating them can make you happier and feel more connected. How does your fast-paced life cause you to miss out on the beauty of life around you? Do you drive fast to beat the traffic lights, walk fast to get to the front of the line, or check your mobile phone during conversations? What are you missing out on with all your rushing around?

Explore

Take a moment to notice how you interact with the world. Notice how you drive, walk, interact with others, and even breathe. Are you always rushing? Prompt yourself or set a reminder several times per day to pause, reflect, and enjoy life, to enjoy Nature. When you hike to the top of a mountain, your story is all about how you got there. Great stories come from your own personal, emotional journeys. They tell of chances, choices, discoveries, and transformations as you ventured out into the world.

Go outside and sit on your steps, balcony, lawn, or open a window. Take the time to pause, breathe, and take in Nature's beauty. Describe how the breeze feels as it plays over your face. Name at least five things that you see, five things you smell, five things you feel, and five things you hear. Perhaps you can even taste something of Nature in this moment of purposeful attention. Find a flower, put your nose up to it, and breathe deeply. Do you discover that you are more open and alive to yourself, more alive to the natural world around you, or curious to what subtle invitations lie in wait to be noticed?

iRewild

Take a couple of deep breaths, slow down, and spend three minutes imagining the rose's fragrance living within you and finding its home within your beingness. How do you feel, knowing that the rose's blissful fragrance now lives within you and resides in your beingness, forever?

Reflection

Week 27

Imagine

In the early spring morning, a small flock of birds wing through the air, their flight setting into motion lasting sounds. The vibrations of their flight, of their songs, seem to stitch together unseen threads of light within the moving currents of air—threads that echo the universe's living essence across the fabric of life. Each individual bird is a unique instrument in Earth's cosmic orchestra. Earth's voice, Earth's wishes, Earth's dreams, Earth's creativity carried upon their songs.

Question

Listen deeply for all the dimensions of the birds around you. What sense of beauty, feeling of well-being, comes over you as you hear the birds fill the rising warm air with their music, singing just for your heart?

Explore

Step outside or open a window. Breathe. Engage your senses and let the soothing and vibrant sounds of Nature resonate deeply within you. Acknowledge the interflow between you and Nature, between you and the birds, between you and the soul of the world. Let yourself become part of this symphony. Breathe. Hear the birds' tender songs as they invite your soul to take flight in the vast expanse of their cosmic choir. Visualize your soul radiating with light, then allow it to ascend upwards and soar on the notes of their sweet, unforgettable melodies. Give your soul time to sing along with the birds, ride the wind currents, to explore, to play, to enjoy.

iRewild

What personal commitment to help Earth will you whisper to the birds to carry back upon their songs to Mother Earth? In 100 years, in 130 years, what story will your descendants tell about the actions you took all those years ago?

It is a truly extraordinary time to be alive since we possess the genuine capacity to make a meaningful difference in preserving our planet. Science is clear that the future is still in our hands. In this important time, what we do, even the smallest gestures and actions, can have a profound significance.

Reflection

Week 28

Imagine

A majestic old oak tree, with twisted branches, rugged scars, and a mossy blanket, seems to call to you as you walk by. You see beauty in his seeming imperfections, strength in his challenges, and you wonder at the wisdom that this magnificent being has gained with age. You get closer, touching his bark. The bark's rough surface is alive with insects, tiny fungi, and lichen—a paradise world to many small beings. You peer into a mysterious and dark arched doorway at the base of the oak's trunk and smile at the possibilities beyond. In the emerald-green canopy, you hear soulful bird calls. You imagine they're singing to the clouds.

As you stand, your body aligned to his trunk, you feel like you two are becoming one, merging into the being of each other. Breathing in the tree's oxygen, breathing out your carbon dioxide, in this moment of oneness, you ponder your own, and all of humanity's, dependence on trees for life. Now you imagine the world beneath your feet, where long roots are laden with mycelia, sending messages to other trees and protecting the earth from erosion. With your connection to this ancient being you rekindle an old friendship, as old as time itself.

You enter a deep part of your soul and heart, and you wonder if there was ever a time when trees flourished in partnership and peace with people. Your mind wanders freely, exploring the possibilities that may have existed. Did we sing to the trees like birds do, or dance hypnotically in a warm sunset like butterflies and gnats in summer? What a kinship and interconnection that must have been! When did our relationship change?

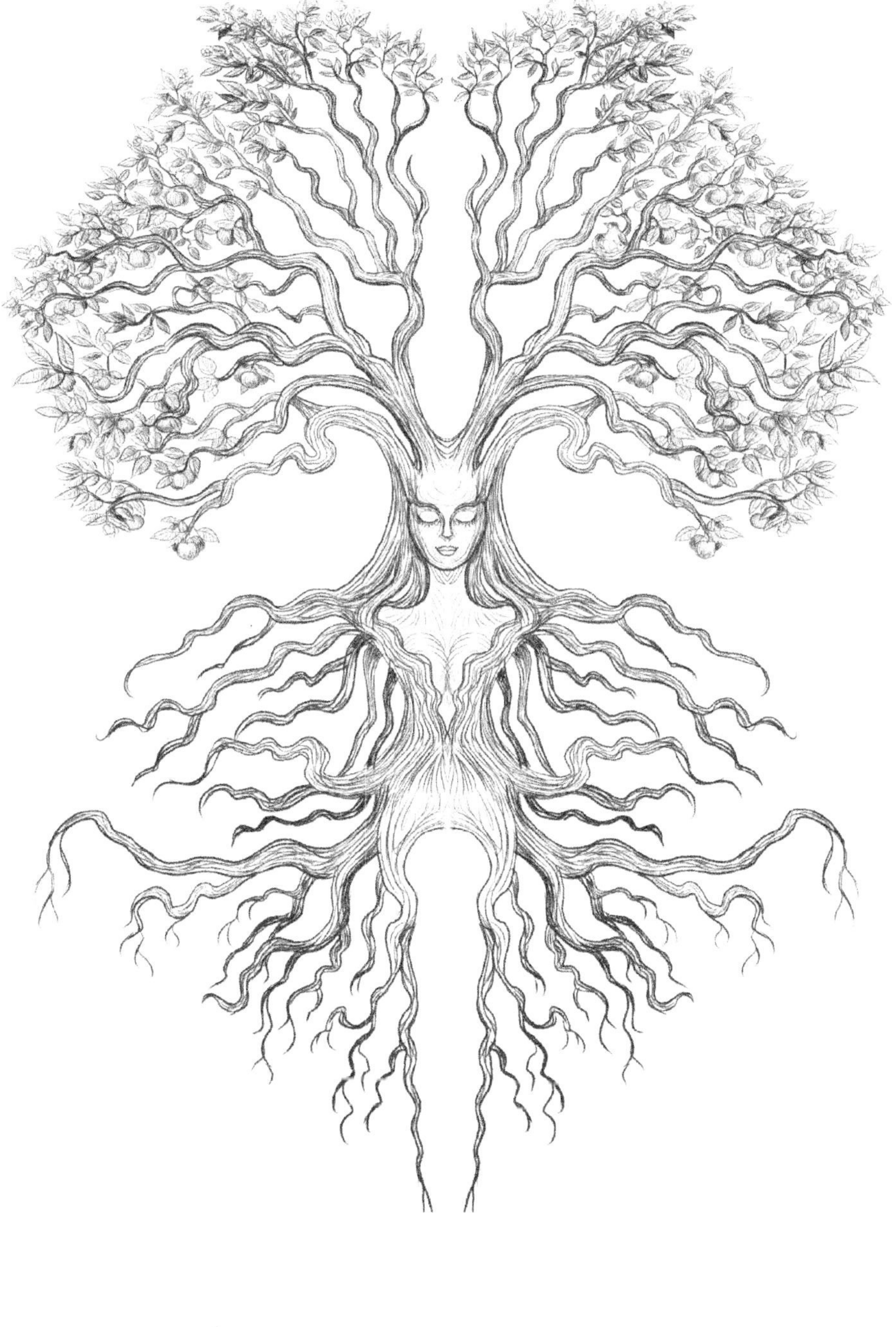

As you walk away, you feel a greater sense of awe and respect for your tree kin. You drop old perceptions of your place in Nature, like a tree's leaves fading away in autumn. You wait patiently with heartfelt curiosity, like a bud eager to spring open with a new worldview. Soon you will return to your kin; this is only the beginning of your relationship.

Question

When you walk up to an ancient tree, what does it think about you and who you've become?

Explore

Now it's your chance to meet and commune with your own tree kin. Go outside and find a friendly aged tree that you are drawn to. Once you've made your introduction, place your hands on the tree's bark and feel its bumpy surface. It's time to build a relationship and deeper connection. Look at the mysterious tiny details in every nook and cranny. See the light filtered through the tree's leaves. Allow the radiance to touch your skin and bathe your body in eternal energy, trickling light flowing harmoniously through your core and the tree's. You are one with your tree kin. What can you sense? What images enter your mind's eye?

Once you feel closer to this great being, maybe the same day, or another, share an interesting story about yourself: a challenge you've faced and overcome; an adventure you've had; someone you've helped; a dream; or a unique place you've visited. Once you finish, give the tree permission to share your story with other trees in the area, and thank it for its life source and companionship.

This tree has stood still, witnessing passers-by and their actions, emotions, and conversations, for such a long time. A sleepy farmer, soldier, or rider may have taken their rest on its solid trunk many years ago. Imagine the eventful times in history through which this mature, humble tree has lived. Maybe this wise elder would like to share its own memories and wisdom with a curious, open-minded soul like you. Sit back and tuck yourself into the tree's embrace, in a caress that feels like you're being held by a loving grandparent. Drop your gaze and invite the tree to share a story with you. Listen through your heart.

iRewild

Like the tree cares for many beings, what species or part of Nature relies on you for survival? Close your eyes and picture that image in your mind.

Like the butterflies dance away a summer afternoon weaving through tree trunks, or a child skips around a waterfall of weeping willow branches, how could you joyfully rekindle your relationship with trees?

Reflection

Week 29

Imagine
In the furthest depths of the mysterious waters that cover the surface of our exquisite planet, millions of different animals live out their lives. Whales the size of double-decker buses dance gracefully together in their family pods, reveling in the sheer joy of being alive, while tiny phytoplankton harvest the energy of the sun and convert it to their tissues, forming the basic, fundamental food chain of the ocean.

Question
Imagine yourself as a sea-living creature (jellyfish, seal, octopus, sea slug). As a sea-living creature, what is your experience of your natural habitat? Has it changed over time? ship with water? Can you write a small prayer and offer this, in gratitude, to water?

Explore
Move your attention into your heart space. Now begin to breathe the ocean deeply into your heart. Feel its waves merging with the beat of your own heart. As you breathe out, breathe yourself back into the ocean from your heart space. In this mutual exchange, you open yourself up to the possibility of merging with the ocean, as you become one. Stay in this space for as long as feels comfortable.

iRewild

As you go about your day, consider just how much water features within it, from washing and bathing, to making a drink, to the rain, to water flowing, in rivers and streams, out to the ocean. The water that the dinosaurs once drank is flowing around, within, and through us, in every moment of every day. How does this knowing change your relationship with water? Can you write a small prayer and offer this, in gratitude, to water?

Reflection

_____

_____

_____

_____

_____

Week 30

Imagine

The mountains are usually the last to welcome spring. But now the cold, barrenness, and solitude of winter have faded. The forests are awash with youthful light green colors, opening their arms to Nature's new visitors. This is a season of emergence, full of promise and new beginnings as life starts all over again. You are invited into an open-air living symphony with the gobbling of young turkeys, a festival of bird songs, the aerobatic feats of hummingbirds, and butterflies bouncing through the breeze from flower to flower, as a rivulet rushes over the rocks, dropping down the mountain alongside you. Mother Earth, the silent maestro, intimately familiar with every forest creature and plant, makes the performance come alive, guiding all the interactions of life in an ever-changing grand concert to the landscape. All these beautiful Earth-born beings animated by Nature, having made their individual journeys from afar to this place.

The voyages undertaken by all these beings are truly wondrous. Imagine what it is like to be a painted lady butterfly migrating 7,500 miles, navigating the Sahara Desert, across the Mediterranean Sea, to reach Europe, using only the Earth's invisible magnetic field for navigation. Or a ruby-throated hummingbird, weighing roughly the same as a penny, traveling 4,000 miles from Costa Rica to Canada, orienting herself by this same intriguing magnetic force. Imagine what it is like to be a hatchling Atlantic sea turtle emerging from an underground nest, without parents for guidance, scrambling into the deep, open waters to begin a transatlantic swim to the Sargasso Sea, where he will feed and grow, and then returning, decades later, to the exact beach where he was born, using his innate, geomagnetic field sensors as a compass. Or a salmon migrating thousands of miles out of the ocean and back to her stream of origin using the Earth's magnetic field like a map.

Question

Science now has evidence that the human brain can respond to the Earth's magnetic fields on an unconscious level, out of our immediate awareness. Have you forgotten, silenced, or dismantled your ability to navigate back to where you came from, somewhere beyond yourself, somewhere you belong? Is it time to set off on your own long journey, back to the soul of the world, to reclaim your place of belonging, your living bond with Nature's intelligence? How do you use your natural compass, your sensory modality that once guided your distant ancestors, to navigate where you belong on Earth—a place where you fully participate with the living world, a creative partnership that restores the very fabric of your soul?

Explore

Not everything in this world can be fully explained, yet you welcome the footprint of its deeper mystery. Engage with curiosity and wonder. Can you see yourself from the viewpoint of a star, a flower, a rock, or the wind? What might your surroundings look like? The forested mountains of Costa Rica? The tropical grasslands of the African savanna? A strange and unique ocean covered in dense seaweed? The gravel beds in the upper reaches of rivers? Take a moment to reimagine and unite with this captivating place. Surrender to the sensory symphony of the landscape enveloping you.

Then, draw or describe in intimate detail what your place of origin looks like. What colors live there? What sounds and smells exist within this landscape? What animals, plants, or other beings reside there? Where are you within this living web of life? How do these beings speak to your soul? How do you call back to them? What does it feel like, what emotions rise up as you experience your natal environment? What surprises you about your surroundings? What seizes your childlike curiosity about this living reality? Explore it from all angles. Capture all of these questions in your description or drawing. Leave out nothing that may seem small or insignificant, as it may be meaningful in future explorations.

iRewild

Take hold of and embody your responses to the exercises. At a deep level of human experience, how are your geomagnetic sensors trying to point the mysterious depths of your own being back to the creative intelligence that exists deep within the heart of the natural world around you?

Reflection

Week 31

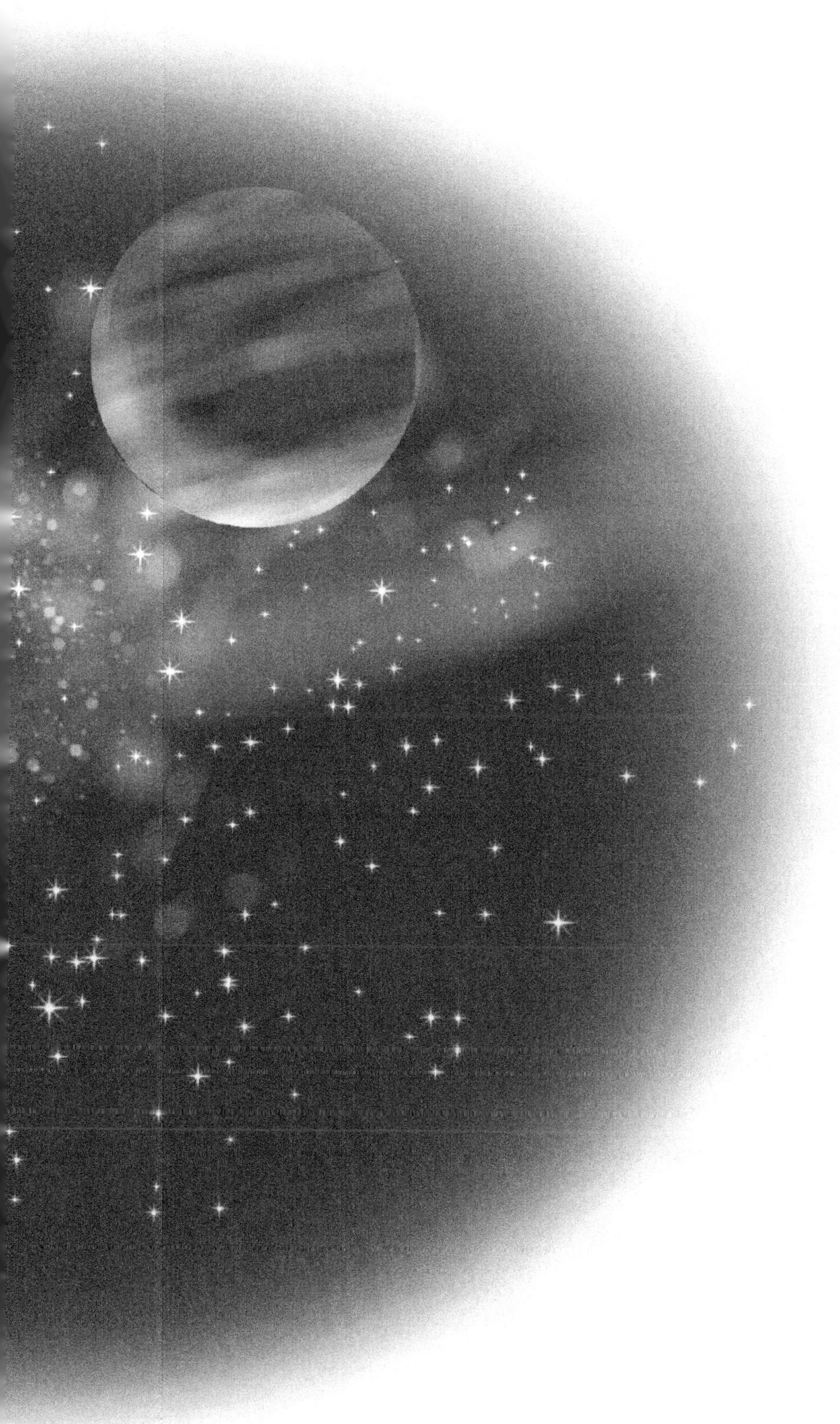

Imagine
There is a frog pond deep in the forest where the flow of time stands still, permeated and embodied by the timeless. When walking, it seems to be close to the ends of the Earth; maybe you have seen it in your daydreams and dreamt of being swept away in its mysteries. The dark pond is ringed by steep, muddy shores, moss-covered rotting logs, and knee-high swamp grasses. Large water lily leaves float across the water's surface and irises blur the pond's edges. Their deep blue petals gracefully contrast the lengthy green leaves. Water spiders rapidly skip across the pond's surface, while fireflies dance on the edge of the forest. Their threads of light become lanterns in the descending darkness. If you stand quietly, the forest life, its soul, reveals itself, but it requires that you be very, very still, silent, and in solitude. Pay close attention.

When the sun begins to set behind the far mountains and the sky gradually turns to a blue violet, the frogs fill the forest's warm spring breeze with never-ending songs they have soulfully shared with all beings throughout Earth's long history. They can't even remember when it all began. An ancient conversation between frogs and Mother Earth, beings more ancient than humanity. The green frogs croon a throaty gu-gu-gunk-gunk; the bull frogs' deep baritone voices add a stuttering rumm . . .vruuum, vruum; all accentuated by the melodious trill of crickets and the distant hoo, hoo-hoo of a great horned owl patiently biding his time for darkness to arrive. Other frogs join the chorus. As one calls, the others respond, elevating the soundscape into a continuous drifting harmony, a lyrical unity bound together by Earth's love and generosity. Then suddenly, it all falls silent. Only to begin all over once again, throughout the moonlit night.

Question
We sometimes experience moments in life when time seems to stand still, moments when we find ourselves spellbound, possessed by breathtaking awe, open to the whole human experience, and eventually transformed. How does the experience of witnessing the beauty of Nature, the interflow, the interconnectedness of life, affect you?

Explore
In an instant, we are born into this world complete, and yet, the creation of our hearts requires a lifetime. Every moment, every experience contains the material for your heart to unfold, expand, and grow deeper, always in new and unexpected ways.

Take a walk in Nature. Find that stillness of body, that stillness of mind, so that a different consciousness can come forward, in which we begin to see the world through a different lens. Listen to Nature, observe the natural world through curiosity, and use all your senses to perceive the landscape surrounding you. Imagine your senses as a doorway to your soul, a doorway to the soul of the world. Notice sights and sounds that your heart hungers to be exposed to, a deep longing reaching out for the divine in Nature.

See Nature's beauty through your different emotions of surprise, amusement, and trust. Find unique value in everything you notice. Look at all the things around you as gifts from Mother Earth—gifts of deep love that embrace you, impregnate you, that tattoo themselves on your heart. Or look at things in your surroundings as an invitation into recognition and belonging; an invitation into a relationship never limited by separation or distance; an invitation into a sense of wonder of the unknowable, mysterious, and divine order and beauty of Nature; an invitation to commingle in a timeless moment with the sacred in the presence of Earth's soul.

These moments bring us to a standstill, move our hearts, and transform us. In these moments when we lose ourselves, when we come into rhythm with our universe, we thrive and evolve. Remember, at every moment, you can choose how to experience your world. You are your world's custodian, the protector of your inner world. You have the power to shape your own experience in this world.

iRewild

How can you maintain that experience of awe and wonder felt in those passing glimpses of radiant illumination? How do you hold onto that connection to the infinite?

Reflection

Week 32

Imagine

In the quiet hours just before darkness, fallen leaves cover the forest path in an array of colors. A snail rests on a red sugar maple leaf, attentively working to break down the foliage and add nutrients to the soil, supporting life in her surrounding world. Her spiral shell is a mirror of our galaxy, a beautiful spiral of stars. She echoes an awareness of our life force and our one, soft body, connected in unity with our home, the wholeness of the universe, a link to both divine and cosmic energies.

The snail is a micro-cosmic symbol, a spiraling path from your outer world, your outer consciousness, inward, to your inner soul. Like her shell that expands as she grows, so does your soul endlessly evolve as you explore inward.

Question

What does your soul long for that keeps you returning to your inner world? How do you respond to it? Do you long to heal the separation between your soul and the soul of the world, our home?

Explore

Spirals are a large part of the natural world—a whirlpool in water, the petals of a pinecone, a wave in the ocean, or your fingerprint. Where do you see spirals? Look out into Nature and find your own meaningful spiral.

iRewild

In the future, how might you look at a snail differently?

Reflection

Week 33

Imagine

The wind travels far and wide, changing direction and speed. Just like emotions and life's experiences, it can bring calm or chaos, joy or sadness. The wind is woven into stories of folklore and superstition, said to bring messages connecting lands afar. It changes landscapes, intricately carves rocks, heats and cools the Earth, and playfully affects ocean currents. It joyfully moves boats and provides energy. The wind disperses seeds over great distances and fertilizes land. With ease, it carries dust and sand from the Sahara Desert that can be admired in an ochre sunset somewhere far away, across the world. Just imagine, the wind you feel on your face has touched the faces of many people before you, in places unfamiliar to your own, with wildlife and tribes you could only read about in books. Through the wind you are connected.

Question

There's only one place on Earth where the wind rarely blows, the "doldrums," a place of stillness near the equator where sailors can get stranded and boats marooned. Like the ships standing still in this windless place, where do you feel stuck in your life right now?

Explore

Play with the wind, like it does with your flowing hair and clothes on a breezy day. Imagine you are both in euphoric sync. Fly a homemade kite, spin around like a whirlwind, sing or dance in harmony with the wind. Whisper words that you'd like to release into the veins of a dry leaf or seed pod and drop it into the wind's passage. Let the wind blow away these thoughts and feelings you have been holding onto.

iRewild

If the deeply Earth-connected people of the Amazon rainforest could send you a message on the wind to help you build your relationship with Nature, what do you think they would say to you? Maybe you'll hear the message crinkling through the leaves of a wind-swept tree. On the returning wind, sometime later, what message would you send about your commitment to Nature? Why not share it one windy day, like you're whispering into the ear of Mother Earth?

Reflection

Week 34

Imagine

A white-tailed deer grazes in a lush field of clover, a tranquil scene of beauty and wonder. Skipping through field and forest, deer live life highly attuned to their surroundings, rotating their ears to listen for the slightest sound, and scenting the wind for whiffs of threat or danger. As humans encroach on wild lands, deer live in an increasingly intense sensory state that changes their behavior and health, and the surrounding ecology.

Question

Fear is the gatekeeper of power—invisible, undefined, and silent. Living in a state of fear is living a life not fully experienced. It penetrates not only our lives, but also the memory of all beings who share this beautiful planet with us.

Has fear prevented you from new experiences? What is the nature of your fears? Do your fears hint at a life of compliance or agreement with the status quo?

Explore

Male deer, also known as stags or bucks, grow a crown of majestic antlers. Like a tree with complex branches, antlers reach outward and upward toward the sky. Draw, or make a collage or hanging mobile, with the form of a branch of antlers as its foundation. From the point of each antler, sketch or attach images of forest animals that symbolize the moral principles and core values that you strive for in your life. Place the artwork where you can see it throughout the day. Then, live your life with these forest animals as your guides.

iRewild

Next time you are outside in wild Nature, bring these forest animal guides into your heart. Ask yourself, are you intruding in any part of their life or wellbeing? How can you use the sensitivity embodied by these graceful animals to help those around you bring more dignity and peace into the lives of an-Other, into the lives of all beings?

Reflection

Week 35

Imagine

Have you ever tripped over a rock while hiking? Often, stones are stepped over and thought of as irrelevant. Like our bones are the foundation of our bodies, rocks form the structure of our planet. There are cultures who believe rocks share a telepathic connection to human souls, while others say they have consciousness, are animate beings, with memories and stories to share. Rocks form the cornerstones of many myths and folklore.

Historically, stones have been sacred to humankind, and they continue to mark our burial sites. Throughout history, rocks have been an integral part of peoples' spirituality and the dwelling place of gods and the spirit of Nature. Rocks have been used as ceremonial objects and conduits for the healing energies of Mother Earth. Meteor stones streak across our night skies as falling stars, and an untold number light up our universe as distant moons and awe-inspiring planets. The life of rocks is endless, a long-living symbol for cohesion, strength, and durability.

Question

You look at the way the winds, storms, and blowing sands formed rocks into different shapes over the ages. You wonder what has gone into the forming of these rocks. What has gone into forming you? As you gaze beyond life's difficulties, you see your potential, you see what is right in your life—character strengths, values, and talents that deserve to be acknowledged. What are the positive traits that you respect, exercise, and make you happy when you put them to use? Strengths of wisdom, knowledge, courage, and curiosity; strengths that nurture a connection to the transcendent, such as awe and wonder, appreciation, optimism, and spirituality.

Explore

Write down a short prayer, poem, or words of gratitude for a stone, then place the paper in your pocket and go for a walk, hike, or bicycle ride. Somewhere along the way, stop and pick up a stone that resonates with you. Does a stone call out to you? Can you sense a being within that stone? Can you capture the essence of the stone? Can the essence of the stone capture you?

Hold the stone in your palms as you whisper the words from your piece of writing. Then, blow a deep breath into the stone. Set the stone back down where you found it, and walk away, knowing that the breath and power of your words are now held within the strength of that stone and have become part of our planet's vision of the future.

iRewild

Throughout history, new discoveries replace old beliefs. For a moment, consider what is not yet known. What if stones have souls? What if rocks are conscious, able to perceive and feel? In what ways would that change your view of Earth?

Our strengths and gifts are a subtle path leading to our responsibilities in life. We carry strengths that the Earth needs. How can you use your strengths to help the Earth's well-being? Like rocks, what other things or people have you considered irrelevant?

Reflection

Week 36

Imagine

Imagine you're walking alone through a deep, dark, emerald-green forest. You hear the branches crack under your feet and the moss softens your steps like a natural carpet. The damp, woody, earthy smells fill your lungs, and shards of light pass through leaves, filling your outer senses with delight. There's a trickling stream in the distance, enchanting you. It feels as though it flows with the rhythm of your heartbeat. The gentle drip, drip, drip of water falls down the side of a shiny wet rock, and the tips of ferns carry it like natural jugs pouring liquid at the edges. It looks like water has fallen here for hundreds of years, maybe thousands. The stone is so smooth, you imagine you could slip down it like a natural slide, into the pool of crystal-clear water below.

Old tree stumps and roots of fallen trees decompose in beautiful, artistic shapes. You even imagine that the decaying trees look somewhat like forest elves, or mystical woodland creatures from local folklore. They're surrounded by new life. Even a wood mouse has made one rotting tree stump its home. Mushrooms line the doorway to its enchanting house, and violets grow on either side. It's so welcoming and homey. You think to yourself that Nature has such a beautiful way of embracing change, death, decay, and impermanence.

Flowers of snow-white, creamy yellow, and cornflower blue raise their heads as you continue to walk. Leaves of all kinds and hues enhance the earthy feel of this intriguing place. You notice dew drops, like tiny crystals, on top of mosses, and wax-like liverworts lining the ground below you. Colorful slime molds and intricate lichen also show their presence, catching your attention as you stroll by. You see how decay is so full of life. Those smells, sights, and sounds are so calming and soothing to your soul. Being in Nature is like a wonder-filled dream, in which you feel true to yourself and part of the natural cycle of life.

How much you've been missing until today! You realize that this is the beginning of your deeper Nature-connection journey.

Question

You feel curious to explore lichens hanging from a deciduous tree; they seem so unfamiliar to you. Their way of life seems so unique, so in sync with the natural world and underpinned by symbiotic connection. The lichen would not exist without the fungus and alga, or the cyanobacterium intertwined, mutually, as one. You ponder, as you wander, on how "at one" you feel with the natural world.

As you walk, you begin to see more decay around you. There's an old white skull of a deer, with small straight antlers, and a freshly fallen young beech tree, its roots still covered in damp earth, its leaves not yet dried through death. In these endings, you see potential for new beginnings, as now you know that without death there is no life—the trees taught you that. You know the deer first fed a hungry carnivore, then carrion-eating birds, and hundreds of small creatures benefited from it; not a single part of it went to waste. Even its sun-bleached skull is now home to tiny creatures and mosses emerging from the earth it lies on. Everything ebbs and flows with the natural cycle of life. No single moment, or life, is wasted. How do you feel about the impermanence of your life? What could you do, or give, that will last way beyond your physical existence? How are you living each precious moment of your time on this Earth? Where are you resisting, or embracing, change in your life? How do you feel about death?

Explore
Go for another walk, this time physically, to a wooded area—the wilder the better. Embrace your imagination together with the wonder of natural reality. Connect to your surroundings through your inner and outer senses. Become part of that habitat like a passing bird that takes a few hours to rest in a safe, welcoming haven. Look for new life in death as you walk. When you find what you're looking for, explore it in detail, without rush. What are you feeling, sensing, and thinking as you explore this natural moment of decay and rebirth? Continue looking for signs of impermanence on your stroll: maybe the flowers are dying, the clouds are moving, the leaves are falling, the seasons are changing, the trees are aging, or an animal lies dead on the ground. How are these moments making you feel?

Now relate your observations to yourself. For example, what limiting beliefs do you need to shed like the leaves on an autumn tree? What new project could you give birth to, like the rotting branch before new life and energy take hold? What makes you bloom out of a season of gloomy light? What emotions can you move through towards reaching a state of calm, like the still, turquoise blue after a stormy, dark-grey sky?

iRewild
Embracing what Nature has taught you during your imaginal and real-life explorations, think of creative ways that you can give back to her, the natural world, to reunite with Nature's cycle of life. If you listen, Nature may whisper questions or ideas to you. We've stepped back, away from our natural responsibility and place in the cycle of life, so ideas may not flow fast into your mind—that's okay. Be kind and patient with yourself as you reconnect your roots with Nature.

Can you now consider how regrowth doesn't happen quickly, but one day it takes your breath away with beauty and abundance? Look at the starry sky one evening, wrapped in a warm, cozy blanket, or step barefoot onto the dry, sun-kissed grass outside, and gently ask Nature to help inspire you to live symbiotically with her!

Reflection

Week 37

Imagine

Death Valley National Park contains many mysteries, including one of Nature's strangest phenomena—rocks that appear to move on their own. One theory is that moisture can make the mud, on which the rocks sit, slippery, making it easier for high winds of over 100 miles per hour to push the rocks along. Another is that these rocks are lifted on ice rafts, making it easier for the rocks to move. Yet another theory is that, as no one has observed these rocks moving, they are moving in another way, a way that, as of yet, is not understood through a contemporary lens. Multiple theories, multiple perspectives; one phenomenon—moving rocks.

Rather than seeking the one true answer, imagine these moving rocks are your teachers. In this sense, allow yourself to contemplate which one of the theories for these moving rocks reaches into your heart and inspires it to sing and dance in relation with life. Allow the rock, not the explanation, to guide you. Just as these rocks are moved, what story of these rocks' movement moves you?

The moving rock's purpose in the unfolding creativity of life is to be a rock that is moved. How it moves, as we can see, is open to interpretation. Perhaps only the rock knows.

Question

What can these moving rocks teach you? What would it be like to sit with a story that truly moves you, beyond explanations and theories? What in the world truly moves you, touches your soul, makes your heart smile, sing, and dance? This is your story. The unfolding of the creativity of life expressing itself through you. Others may look at your life and offer explanations; however, do they really capture your story, your passion, the things that really move you as you dance with life? Have you ever contemplated your life and purpose in this way? How can these moving stones support you to discover who you really are?

Explore

Breathe in and out through your heart for a few minutes, breathing in and out at a depth and rhythm that is right for you. Bring the story of the moving rock into your heart's awareness. Smile, and invite your heart to smile. Allow your heart to create a beautiful image of the rock that moves. How does your heart imagine that the rock moves? Sink deeper and deeper into this growing vision that your heart is offering to you in conversation with the moving rock. Is the rock moving? If so, do you know how? Is it necessary to know how? What might it feel like to not know, but just to observe the rock simply being a rock that moves? What is different for you in this mode of awareness? Could you engage with more of the world in this way? Embody the rock as your teacher. Keep a journal and describe what you discover.

iRewild

Heart awareness is discerning, yet also open to the greater mysteries of the world, to change, deepening, and expansion. Your heart guides you how to come into balance in every living moment as you dance with the unfolding, living, breathing world. Hold the moving rock in your heart's awareness as you navigate your world. What is this being teaching you? Carry the moving rock with you in your heart as you go out into the world, as a living symbol that embodies your dance with the natural laws of balance.

Reflection

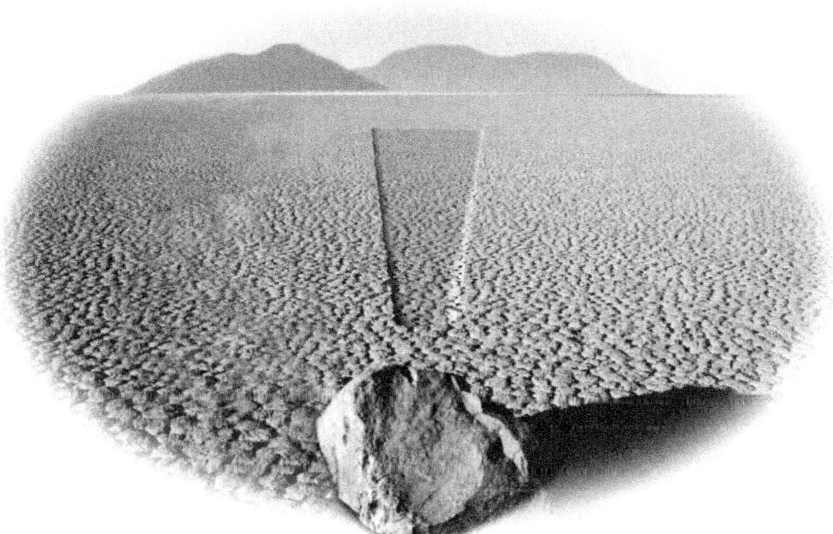

Week 38

Imagine

Fall represents a time for introspection, reconnecting with ourselves and those in our lives. We fine-tune our inner worlds and, by allowing space and silence to take hold of the seeds we plant, create room for new possibilities to grow. It is a time of year to compost old biases and outworn emotions and use that energy to fertilize the soil for stirring the fires of hope, latent seeds of creativity, and greater visions, enabling you to excel and create an ever-better world. Then you rest, moving silently from darkness towards the light, and, at long last, you listen for what wants to emerge.

Nature understands the true extent of cultivating life. In the fall of each year, many plants, flowers, and trees drop their seeds onto the same ground that gave birth to them. Their leaves soon follow, covering and protecting the seeds from winter's snow, ice, and bitter cold, until spring, when the sun's warmth contributes to renewed life.

You are the farmer of your life, the gardener of the seeds of your existence. If you nurture your seeds, if you give them the nutrients, water, warmth, light, and space to grow, their roots grow deep, creating a solid foundation that enables the young seedlings to flourish and new opportunities to emerge.

Question

Are you in an environment that nurtures you and allows you to grow? What arouses your curiosity? What activities inspire excitement or passion in your life? What do you need to add to your garden of life for your roots to grow deeper?

Explore

Using your fingers, dig into the earth. Feel its texture, feel Mother Earth's creative energy in your fingers, hands, and up into your arms. Allow the pulsing energy to flow into your body, expanding your heart. Imagine pulling out any weeds representing old biases and taking them to the compost. In the soil, bury a few seeds from plants that inspire joy within you. Add some water. Each day as you care for your seeds, pause, reflect, and imagine yourself as the spirit of Mother Earth, waking up, providing for the seed's growth, while protecting this tender life during gestation.

iRewild

Sense the aliveness in the world around you. How can you nurture your way of thinking to consider the interests of our natural world? Name three things that you can do to cultivate and grow your relationship with Nature. What tender seedlings are you nurturing?

Reflection

Week 39

Imagine

As day turns to night you find yourself being drawn to a glowing fire. Mesmerized by the flickering flames, the sound of the fire's crackling and popping seems to speak to you in hushed tones...

I am fire, the comforting, soul-soothing element that warms your heart and home. I am the fuel that nurtures life's beginnings, part of the natural ecological balance. I burn away the old and invasive to make space for new generations of wildlife. My grey clouds rain down ashy glitter upon the land, Nature's land, our land, fertilizing and regenerating her soil, and farmers spread my cinders to help their crops thrive.

As my smoke rises, you look for images within it, searching for mystical messages from afar. My crackling sound and fiery colors evoke feelings of security, comfort, belonging, and community. I held your ancestors in my cozy arms for thousands of years as they danced, told stories, and made music around me. Oh, how I miss those traditions. Your myth and folklore revered me through powerful stories: the phoenix, the dragon, the salamander, and all kinds of fire deities from across the world.

I am an architect that creates new land and islands, spewing out of volcanoes from deep within the Earth. I am the product of a lightning strike bursting a tree into flames. Your people juggle with me on beaches after sunset, heat their delicacies at the edge of my flames, and line pathways with me where I burn hypnotically on top of torches. I am the candle on a warm summer's evening, the romantic centerpiece of your garden table. I am dangerous and feared, yes, but, like earth, air, and water, and all the devastation we may bring, I am necessary and revered for life.

Question

Our modern lifestyle may take you away from fire's natural enchantment, but it waits, patiently, to be awoken by the flick of a flint. Do you remember sitting around a fire sharing stories? Warming your hands on the entrancing orange flames? The smell of fire-cooked food? How do you feel in fire's presence?

Where could you bring fire's energy into your life? What part of you, of your life, needs regeneration, like the ashes of a fire feeding your soul?

Explore

Invite fire into your world. Bring it to life in a log fireplace, garden fire pit, campfire, or a candle upon your table. Connect with fire in a way that feels accessible and comfortable for you. Think of how fire sustains our species. Are there any other animals on the planet that need fire like humans do? Look into the flames. What thoughts or inspiration are sparked in you?

As you stare, take your mind back 200 years, 6,000 years, 400,000 years ago, and imagine how your ancestors felt in the light and warmth of fire's protection. Draw images or words that appear in your mind's eye as you watch the golden embers. Perhaps a story or poem comes to mind. Invite others to share fire's comfort, laughing, joking, storytelling, even singing and dancing around its warm light, as humans have done for millennia.

What do you feel in fire's company? What do you sense in your body? What possibilities or burning desires are opening up in you and filling you with energetic fire as you stare into the crimson flames?

iRewild

There are many things humans do without realizing the damage and devastation they cause, just like flames when they become greedy for growth. Alone, or together with friends or family, be inspired by fire's hungry flames. Think of one destructive thing you do to Nature, then write it on a dried leaf or recycled paper. What thing or activity can you extinguish from your life today?

Ignite your new start now, like the new beginnings fire brings about. Throw this one thing you will stop doing into the flames as an eternal declaration towards your part in Nature's cycles. Let fire burn it away and give you strength to hold back your desire to consume.

Reflection

Week 40

Imagine
Some scientists are postulating that the universe is a living organism because of the life cycles that appear within it. In the 1950s, scientists discovered that carbon and oxygen atoms are forged at the end of the life-cycle of a star, dispersed throughout the universe by the star's explosion. All the atoms that make up our human bodies could therefore only have been created by the explosion of a dead star. Perhaps we could say that this is how the star's creativity was dispersed, creating the beautiful world that we live in.

Question
How can you understand yourself in light of this creativity? How might you align yourself with the immensity of what we now know? How can your mind move to appreciate the creative desire of the universe that has been in existence for 13.8 billion years, evolving elements into stones, plants, animals, you, and the whole human race?

Explore
Go out into the world and drop into your heart space. As you move in the world, become aware of Nature endeavoring to claim your attention. Perhaps a particularly exquisite flower speaks to you, or a stone captures your eye. As you sit or stand in relationship with the being that has claimed you, allow yourself to contemplate that you both are created out of star dust, from an initial spark of creativity.

iRewild
Do you feel that anything has changed for you through this exercise? If so, what in particular? How might you act upon this change in the world as you go about your day-to-day life?

Reflection

_____

_____

_____

Week 41

Imagine

Artistic expressions, those we experience and those we create, can be pathways to other connections with Nature. Nature herself, as art, can lead us on this journey of forming new perspectives and connections. Permeating this journey is our evolutionary bond with sight. Even as we utilize our sight to experience art, the artist may be trying to teach us to see without our eyes, to experience without our usual senses.

Art may be helping us access information and pathways beyond what we usually experience. It is here where insight may lie, together with deeper learning, and an extended connection with Nature. Art reimagines our senses and takes us on a journey into our souls to see without eyes, to touch without hands, to feel with a deeper sense, as if to increase our dimensionality. Invocation of a purely imaginative experience may be our journey to deeper insight and connection with Nature.

Question

As sight-dominant beings, we construct our relationship to Nature primarily with light. Did you know that we have many more senses than you may be aware of? Beyond the familiar five senses (sight, hearing, touch, taste, smell), what other senses within you are you already aware of and could expand upon in order to develop a deeper relationship with

Explore

Artistic expression is one way of connecting with Nature, interpreting Nature, and developing an understanding of our place within Nature. Throughout your journey with this book, you have breathed life into your imagination to develop a more conscious relationship with Nature. Now it is time to transform this imaginative experience into something you can not only see in your mind but can touch and see outside of your thoughts. Choose a passage from this book that has touched your heart and give it another life with your creative hands. This is not to diminish your connection to the words or inquiry, but to create another expressive representation of your feelings. See if the process and act of transforming your thoughts into something physical deepens your initial expressive force. What can be learned by taking a deeply personal experience within yourself and then creating something to see, to touch, maybe to hear or even smell, and possibly to share?

iRewild

Using our imaginations to contemplate different relationships with others and the world around us, as well as the world we may never see, feel, or understand, gives us the opportunity to connect more deeply with Nature. With our minds, our hearts, our souls, and our hands, we can create a world that defies conventional boundaries. Can you think of what you have created as part of a bridge you are building to somewhere beyond your existence, beyond your senses, to other connections? Can what you have created reveal something else, something other, something unimagined? Will you let it?

Reflection

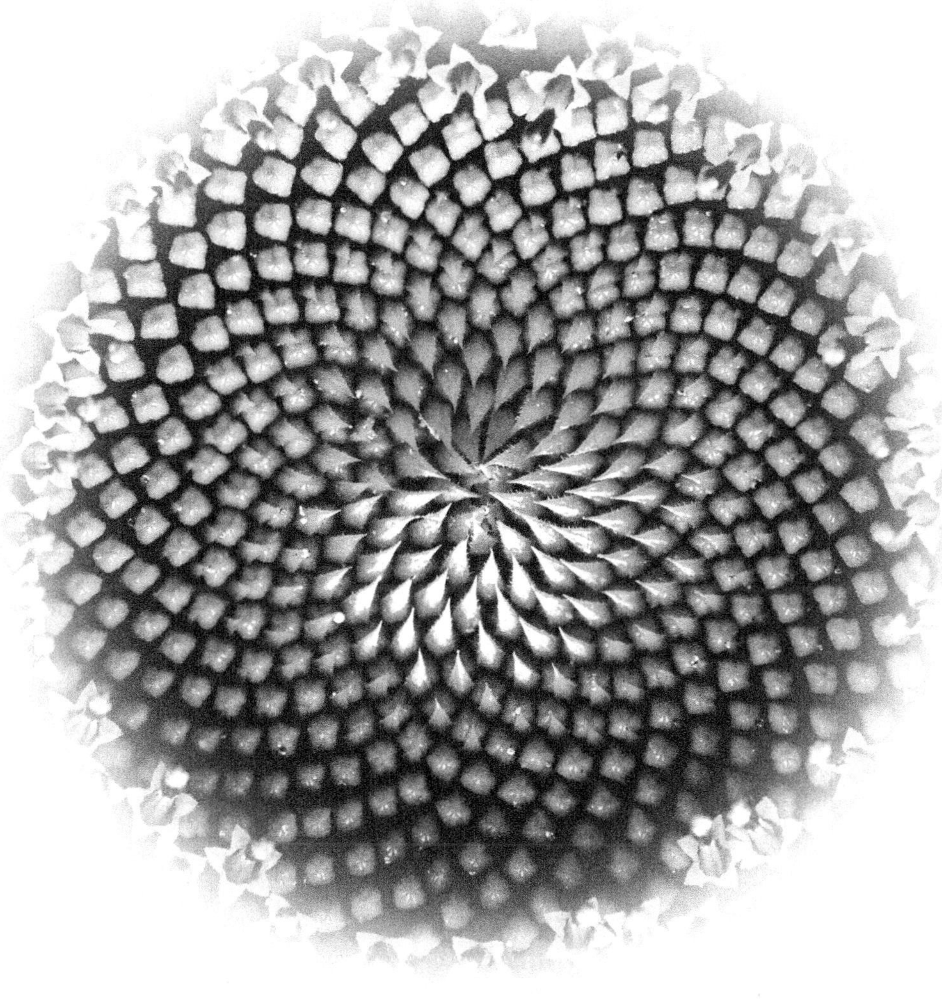

Week 42

Imagine
Fireflies dance on the edge of the forest like guiding lanterns in the surrounding darkness in a celebration of summer, of life, and of Nature. Their threads of light evoke a childlike innocence of when the world was seen as a magical dreamland and nothing was outside the realm of possibility. Fireflies—just a beetle on the outside, and yet, a brilliant light in the life of an-Other.

Question
How often have you felt like a beetle on the outside and found out later that you made a significant difference in the life of an-Other?

Explore
Go outdoors and stand before a sunrise or sunset; stand before a long-fallen tree or a wild blossom as it unfolds; stand before the fields of blue sky or the swelling music of the ocean. Stop. Deeply observe and experience the beauty and wonder. Feel the sun's unfolding touch upon your face as it extends light and warmth, restoring the Earth. Notice the roots of the surrounding trees as they nurture the leaves that give you breath. Touch the damp soil, the legs of a plant, or the diamond-like dewdrops that give us nourishment and beauty. Hear the ceaseless and tender songs of the birds as they invite you to break the barren impressions of yesterday. Observe

the squirrels, bees, mourning doves, and other creatures that help sustain our lives. Linger in this happiness.

Begin each day rooted within this gentleness and unity. This is what prayers of gratitude to the Earth are made of. Take a few moments to write a short prayer of thanks to all of life. A brief prayer that you can express to the world each time you step outside. In can be as brief and powerful as, "Thank you for this beautiful, life-giving world with all its plants and animals, skies and waters, and its invisible wonders and human beings." Now, create your prayer. Then, quietly, with intention, recite your words and let them echo throughout the delicate universe each time you step outside.

iRewild

Can you imagine a world without the light of the firefly? Beauty in Nature inspirits our hearts and minds with awe and wonder, luring us even further into our curiosity and encouraging us to reflect on life's bigger questions. What is the biggest question that Nature needs you to ask yourself?

Reflection

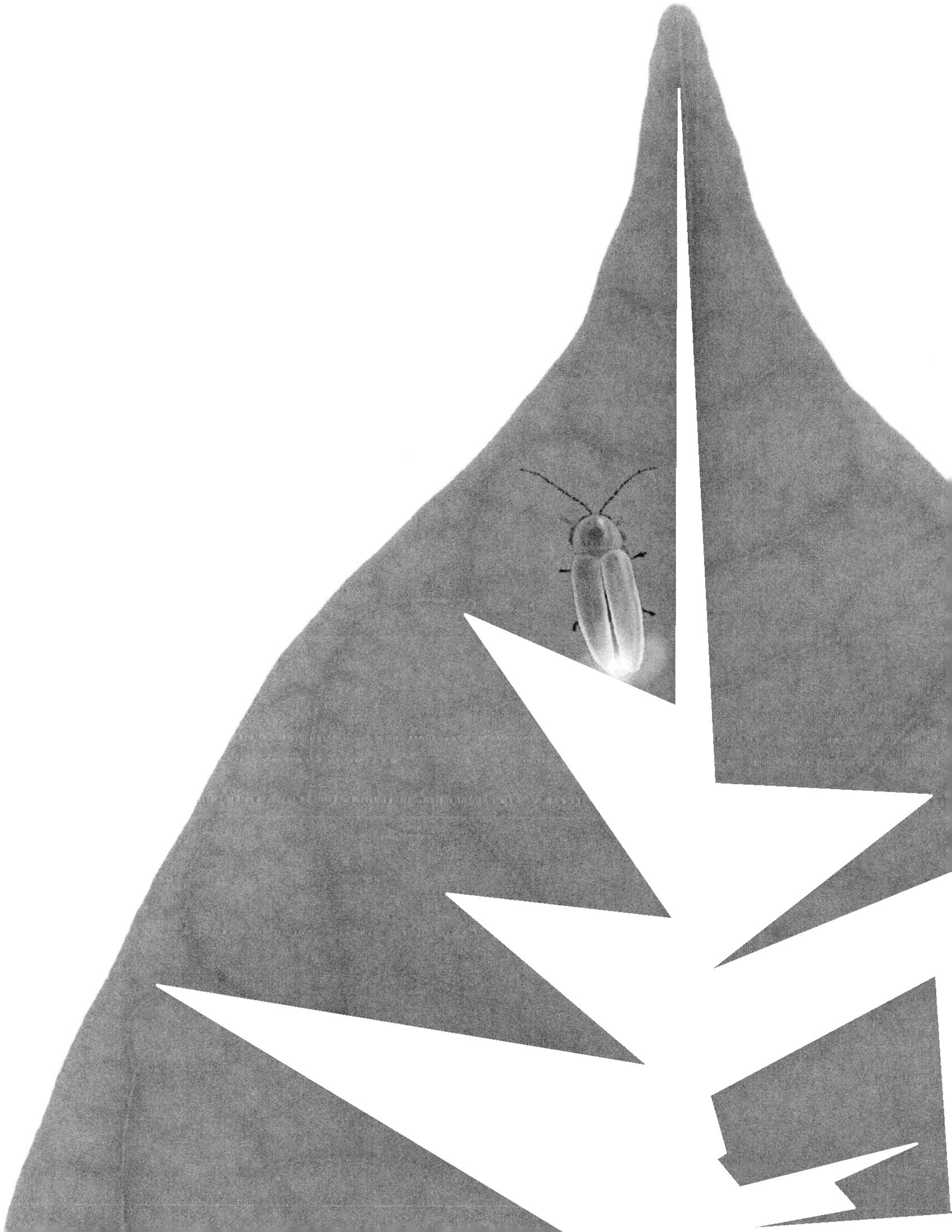

Week 43

Imagine

Purple periwinkles are the first flower to welcome spring, to brave breaking through the mountain's winter embrace on the woodland's edges. Flowers of pure joy, of new life, of new beginnings—they are happy to see you.

Within days, the periwinkles appear to transform into butterflies. The spring azure butterfly, as deep a blue/purple and as small as the periwinkle, dances above the flowers, one a mirror for the other. How did the two find each other? Nature's colors take flight, each celebrating the other, each welcoming the other, each giving and receiving from the other. They dance, weave, and dart about, celebrating Nature, adding to its story as it wakes once again from its deep winter slumber, flourishing into spring.

Question

What does it feel like to dance, weave, and dart about, celebrating Nature, seeing yourself reflected in Mother Earth's story, as you wake once again from your slumber, flourishing into life?

Explore

Does your heart skip a beat when you notice the first butterfly or birds of spring? Step outside your home and notice, really experience, just one bird or one butterfly who has stopped by for food or rest within your surroundings. The land around you offers this being everything it needs to sustain its life and travels before moving onward. Understanding this one being's knowledge, life-ways, and history is integral to our relationship with this being, to our landscape, and to the challenges that this being faces that stem from pesticides, urban development, and depletion of natural resources. Such barriers prevent this being from experiencing the world as its ancestors did.

Research the one butterfly or bird that you noticed. Where does it spend its winters? Is the landscape of its natural habitat mountainous, grasslands, or wetlands? How far does it migrate to reach the area in which you live and what obstacles does it encounter on its journey? Where does it travel to after leaving your home? What types of food and shelter does it need to sustain its life on its travels? What stories does it tell?

What is this being's history? How many millions of years has it existed? Does it have a cultural relevance to the people where it resides or the landscapes it travels through? Are there beliefs, superstitions, or myths that incorporate this being? What has been its history in your area?

How does this being help maintain Mother Earth's ecosystems? How may it and its habitat be threatened? How would you feel if it became extinct?

Witness, recognize, and appreciate that your backyard has been an essential, life-giving place that nourished this being with nectar-rich wildflowers or the perfect place to lay its eggs. Your backyard provided this being a safety net before lifting off for faraway places.

iRewild

They invite you to play, they invite you to notice the beauty in Nature. Whimsey and work. If you were a butterfly working in Nature, what would you choose to do? Forget the burden, the heaviness, of everyday life. For a moment, go from grounded to flight.

Reflection

Week 44

Imagine

You find yourself deep within a forest, sitting next to a mountain stream. You hear the water's gentle sounds burble as it travels along the river bottom, bubbling over rocks and fallen tree branches. As you watch the drops of water in the stream trickle toward you, you sense that it is filled with joy, belonging, and the desire to fulfill its purpose. Rushing drops bump into rocks, rolling and giggling, wanting nothing more than to live on within an-Other—an inseparable part of life, a gift of life to all beings on Earth.

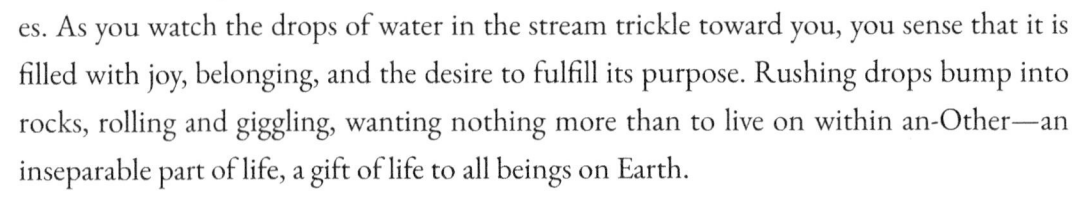

Question

Within the realm of Nature, nothing lives exclusively for its own sake. Nature's essence lies in living for something beyond itself. When you take a refreshing sip of water, can you sense the water embracing its intrinsic purpose as it nourishes and becomes a vital presence within you?

Explore

Stones have memory. Find a stone that captivates your attention and wishes to work with you. If you could make a positive wish for the Earth, what would it be? Create a strong intention and hold that wish in your heart and mind. Hold the stone in the palm of your hand. Take a deep breath and blow that wish into the rock. Find a body of water—a stream, river, lake, marsh, or ocean. One more time, with great intent, blow your wish into the stone and place it into the water so that your wish, your prayer, can be carried out into the rest of the world.

iRewild

Can you perceive the water fulfilling its purpose as it replenishes your body and sustains your being? In the future, how will you consciously consider the significance of each drop of water you drink?

Reflection

Week 45

Imagine

A wave of leaves carried by the wind pirouettes around your ankles as you walk beneath winter's sleeping trees, their past life crunching beneath your feet. You ask the trees, "Do you hear me? When will you awaken? If you're sleeping, who takes care of life, providing the oxygen we all require?"

Through the silent, barren trees the clouds float by, and you sense your life is enveloped in the tree's shadows and fallen leaves. Their naked branches sway with the wind, in relationship with the Earth's spiral dance. You take a deep breath and think of the trees, your friends, your caretakers. Your heart fills with gratitude and awe.

Question

Do you sigh or smile in delight when you've been seen, appreciated, and loved?

Explore

Consciously reconnect yourself to the Earth and the land on which you live. Stand outside in a forest or city park. Feel the aliveness of your surroundings. In your imagination, hear the voices of the land, wildlife, plants, and trees sharing messages of love with you. It's the same deep love that formed your being. Create a strong intention of love toward the Earth, for all beings, both seen and unseen, and hold that wish in your heart and mind. Now pick up a stick or use your finger and draw the word "love" on the ground. If you prefer, bring a lock of your hair to bury in the ground, establishing a relationship with Nature and the Earth. Then, place your hands upon the ground and express your appreciation and thanks to Nature, taking a moment to honor the spirit of the Earth and all the beings that share their existence with you.

iRewild

When you touch the ripples on the tree's bark, can you almost feel the tree sigh with appreciation of being seen, of not being irrelevant?

Reflection

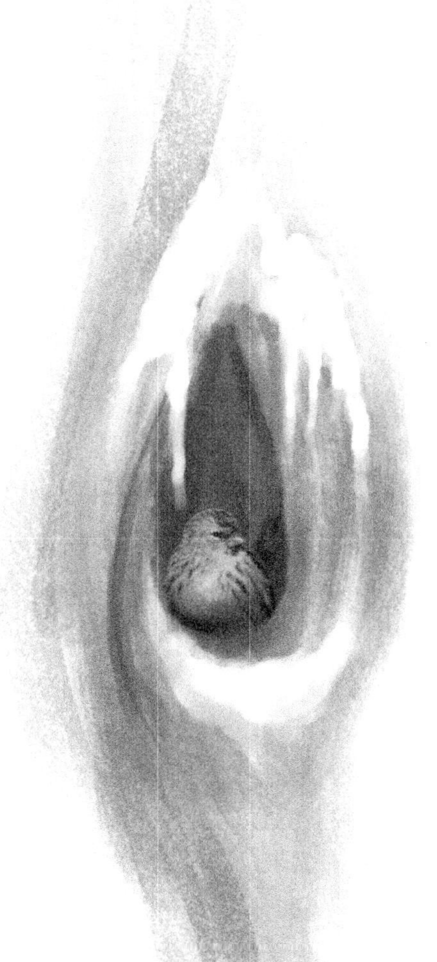

Imagine

A tiny green shoot pushes his way through the cracks in a city sidewalk. Fragile, delicate, yet at the same time, so courageous and powerful. Nature reaches out to us, revealing herself to us, even in the most seemingly denatured places. In a selfless gesture of love, Mother Earth's offspring offer themselves to us, willing us to see, hear, feel, and experience the mesmerizing creativity and ingenuity of life. You look a little closer; a small insect crawls over a wall, going about her daily business—where is she going? Is she going home to her family? Your senses become sharper, as you allow yourself to become aware of more and more: the sound of a bird in the far distance; the feel of the wind on your face; raindrops falling out of a darkening sky. Even in the city, you realize that Nature is all around you if you just stop and wonder about the life that is teeming amongst the concrete and towering buildings. Even in this place, Nature is always speaking to you, reaching out to you.

Question

From one perspective, it's apparent that humanity has encroached more and more on Nature. However, just as the tiny green shoot pushes his way through a crack in the sidewalk, Nature is always searching for a way to return. Think of the countless abandoned buildings, train stations, industrial units, and factories across the world that, over the years, Nature has reclaimed, thriving in places where humans can't, providing a home to a wide range of wildlife, flora and fauna. If you were an abandoned building or a concrete sidewalk, how might Nature reclaim you? What might happen if you gave yourself permission to take Nature's hand, even in these seemingly neglected places? What might you discover? What might you learn?

Week 46

Explore

Place your palms over your heart and breathe into and out of your heart for a few minutes. When you feel ready, walk out into the world and make a silent request for somewhere or something to make itself known to you where you can sit and connect—this could be a plant, or a tree, or a small, green space. Let yourself be drawn forward. Once you have been called, sit down, close your eyes, and commit yourself to connecting to your "sit spot," a place where, even in the midst of a town or a city, you can connect to Nature and hear her voice, showing you that Nature always finds a way and is always reaching out to you.

iRewild

Contemplate this statement: even in the city and in the most abandoned of places, Nature's presence exists as a background force of creativity, out of which all arises. How can your reflections on this statement support you in engaging more deeply with the Nature that surrounds you in your day-to-day life?

Reflection

_____

_____

_____

_____

_____

Week 47

Imagine

Long, long ago—before your time, before your great-great grandmother's time—there lived an ancient yew tree. This magnificent tree watched the passing of the seasons from the sacred land on which she stood, year by year, century by century, and, so very slowly, put down her roots and reached her dark green crown to the sky. This tree originated and lived out her life in deep time, over 200 million years ago. As the Earth turned and the years unfolded, countless changes occurred across our precious Earth, including five great extinction events. Yet the yew lived on, carrying the seed of the modern yew trees that live in churchyards across Europe and in quiet, countryside hedgerows. Indeed, there are many yew trees currently in existence that are as much as 5,000-years-old—just little saplings when the great pyramids were being constructed in Egypt.

Question

Contemporary science tells us that modern humans have only existed for approximately 300,000 years. Compare this with the 200-million-year lifespan of the yew tree species. How does this make you feel? Can you contemplate what the yew tree's experience of the world was like all those millions of years ago? What can she teach us about her strength and fortitude through difficult times? What can she teach us about humankind?

Explore

Imagine an ancient yew tree in your heart and mind. Stay in this space of contemplation for a while and listen to what your yew tree ancestor wishes to say to you in relation to modern times.

iRewild

How does it make you feel to realize that our

current view of the world, which has been in place only for the past several hundred years, has led us to become disconnected from Nature? Can the yew help you reconnect? Can you see the world through her eyes? How can you become an ancestor worth descending from, like the yew? How does this make you think about your own ancestors? Make some notes about how you might like to be in the world differently, with the support of your ancient yew tree ancestor, and your own ancestral line.

Reflection

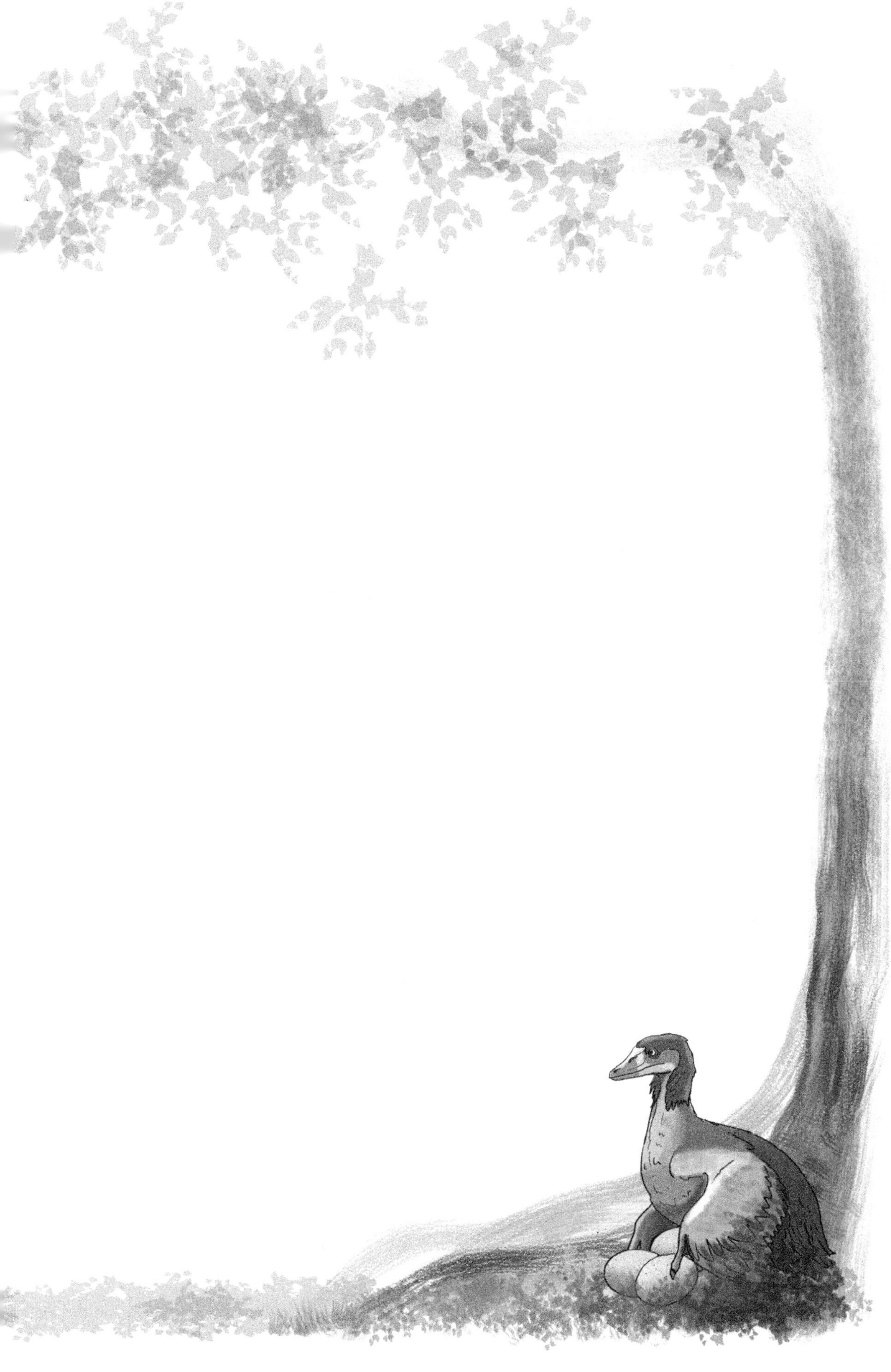

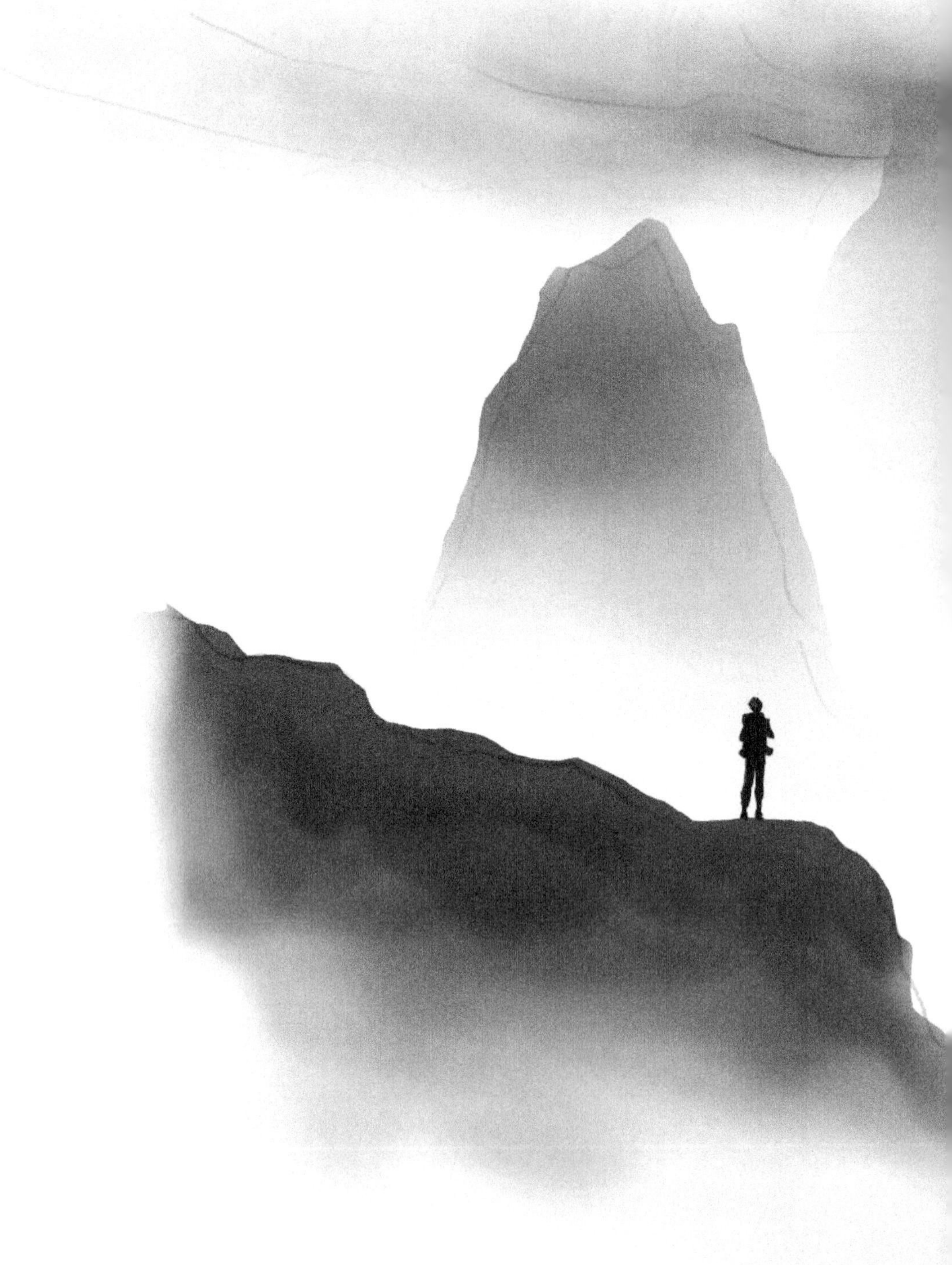

Week 48

Imagine

Woven between the fabric of modernity and a land imbibed with myth, the towering, weather-sculpted stone mountains rise above the valley floor. In the light of dawn they stand in solitude, saturated in reds, golds, and corals, engulfed by vast deep blue skies and billowing clouds. You stand in awe and gratitude as you take in the towering monoliths and grand spires. Some of them soar a thousand feet above you, turning the commonplace into wonder and mystery, luring you even further into your curiosity and encouraging you to reflect on the bigger questions of existence. Your heart fills with profound appreciation as you open your awareness to your surroundings, to the rich history, myths, intricacies, and value of this magnificent world and its future. You are reminded of your kinship with Nature, her gifts that sustain your life, and that you are an interconnected part of all life on Earth.

Question

Imagine if you encountered your life experiences in such a way that the first thing you saw was beauty. Beauty before judgment; beauty before opinion; bevauty before criticism; beauty before expectations; beauty before reacting. How would that change your life experience? How would that change your enjoyment of life and the ability to appreciate its many gifts?

Explore

The power of gratitude shapes reality. It enlivens and cultivates a more profound connection between your heart and our natural world. The spirit of appreciation builds bonds that help to nurture love for our planet. Deepen your relationship with all the elements surrounding you by giving them thanks.

Go for a walk or drive to a natural area where its beauty fills your heart. It could be a forest or a garden filled with flowers. Stop and take a slow, deep breath, and gradually notice things, admire the beauty that Mother Earth has given us, beauty that has gone unseen. Close your eyes and take another deep, full breath, filling your lungs, and slowly exhale as you smile. Reopen your eyes and notice the subtle beauty in the sun's rays. Sense the cool breeze on your skin. Notice the vivid colors around you and the repeating patterns of Nature. Open your whole being to the aliveness in all the elements of the natural world around you. Take the time to appreciate it, to allow the beauty of our world to fill your h heart. Let those feelings spread throughout your entire body. These brief experiences make us happier, give us energy to take on the day, and help define who we are. Within these fleeting moments life takes on new meanings.

iRewild

Cut ten small strips of paper. On each piece of paper write down one thing from your Nature-experience that filled your heart with beauty and gratitude. Place all the papers in a transparent jar. Each day this month, add one more gratitude. At the end of the month, open the jar and reflect on the notes through the lens of gratitude. From that deep place within, what emotions and thoughts do you experience when you think about Nature?

Reflection

Week 49

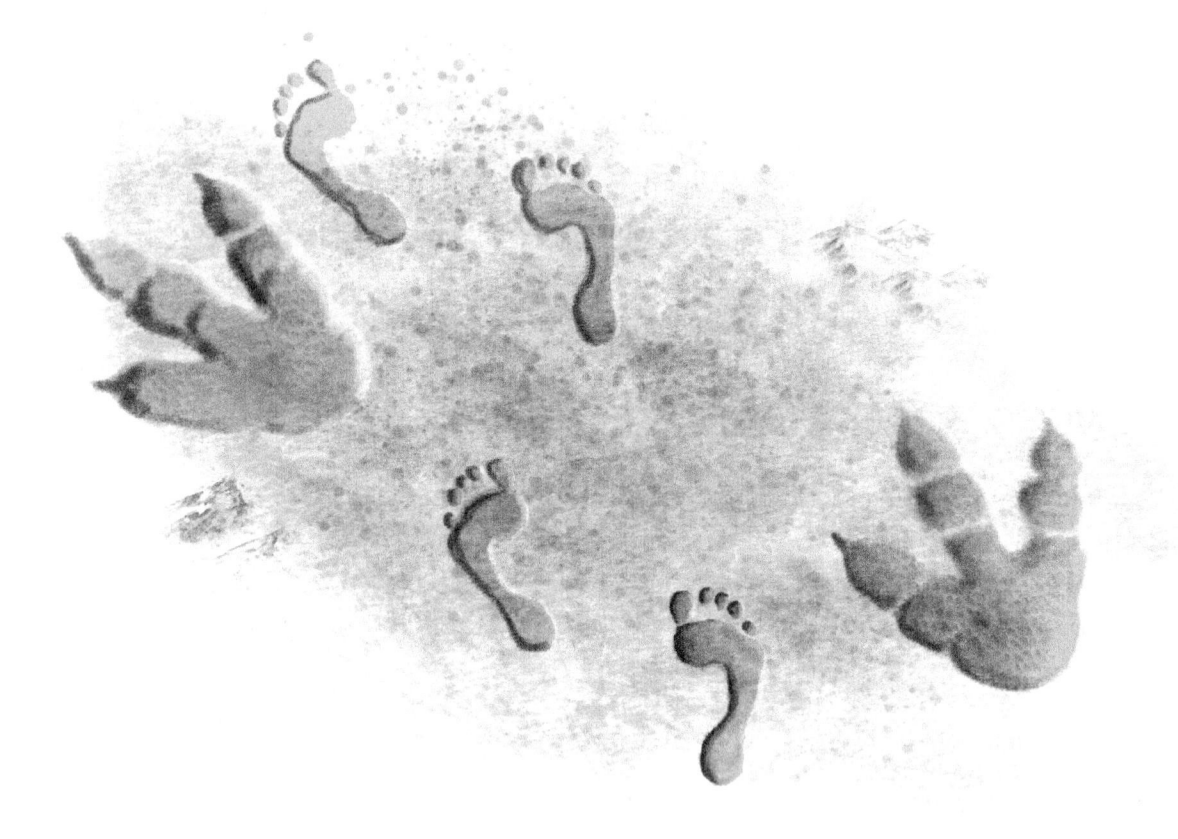

Imagine
We have inherited this living, breathing, beautiful world from our ancestors. Generations upon generations lived before us through a myriad of very different time periods. All these humans experienced a relationship with the Earth, in many different ways, and, in doing so, helped to create the world that we live in today.

Question
Just as you are contemplating the lives of those who lived before you, can you now imagine yourself in your great-grandchildren's world? What would you wish the world to be like for them? If your grandchildren were by your side now, how would they see you interacting with Nature?

Explore

Moving your attention into your heart space, breathe the world deeply into your heart. Feel the world touching your heart, offering you its unconditional love. As you breathe out, breathe yourself back into the world. Feel this participatory relationship growing in your heart space. Can you feel the hearts of your ancestors reaching out to you through time? Are you able to sense your own heart reaching forward in time to the future inheritors of this world? What arises for you as you do this? Continue this relational breathing for as long as feels comfortable.

iRewild

Has this exercise changed you? If so, how? Make notes or create a piece of artwork. Is there anyone close to you with whom you feel comfortable sharing your experience? If so, how could you share this with them? What do you feel would be the importance in doing this?

Reflection

Week 50

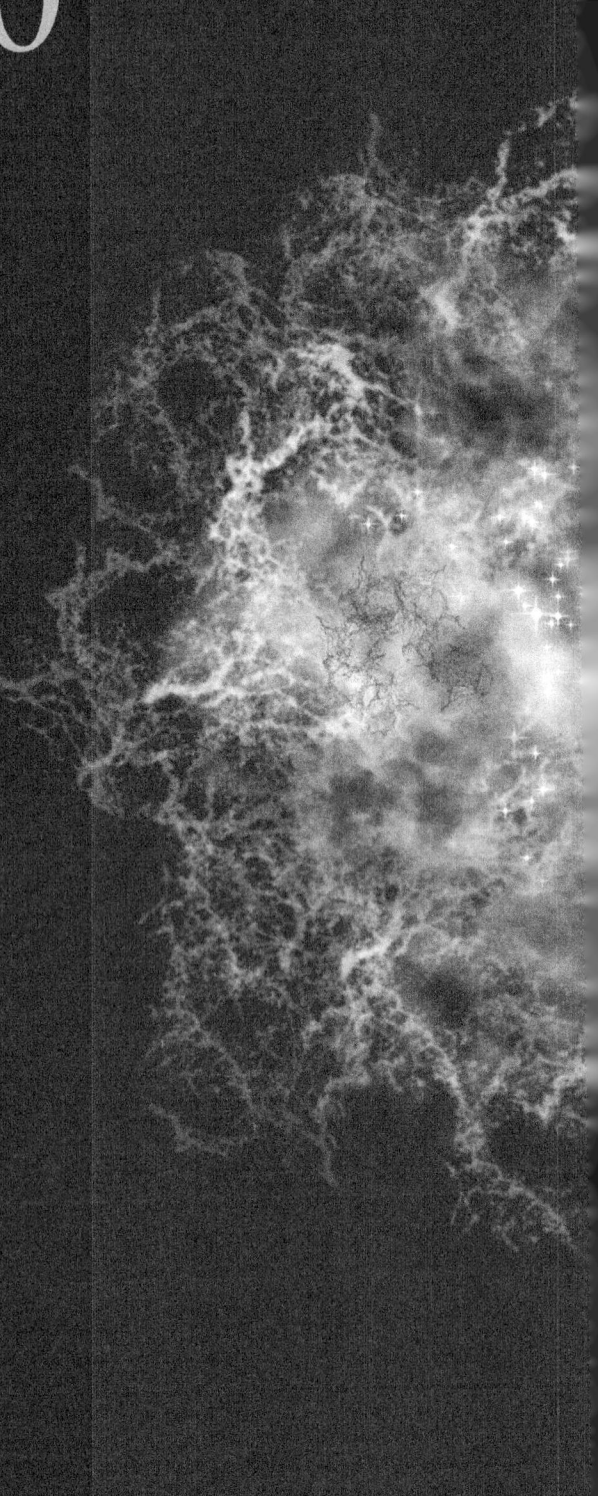

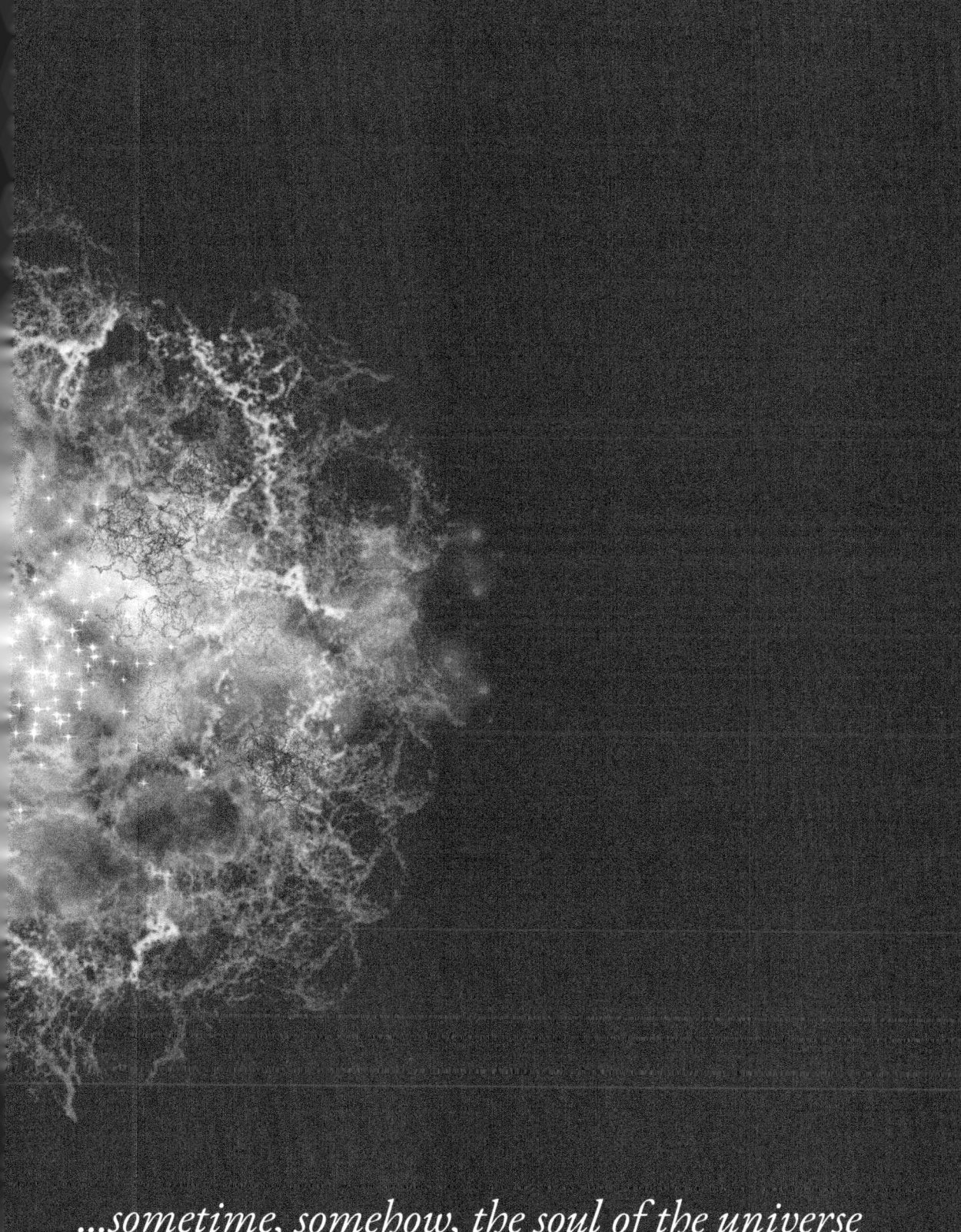

*...sometime, somehow, the soul of the universe
took a great exhalation, a cosmic breath...*

Imagine

Somewhere, sometime, somehow, the soul of the universe took a great exhalation, a cosmic breath, that continues to the present moment. Before she breathed, she was pure beingness, without form, time, or space. Then from that beingness, her consciousness, she gave birth to our everything—our time, our space, our creation. In that breath, all that is was born and continues to come into existence everywhere, at all times. With that exhalation, all of the universe and beyond seems to have fulfilled a primal imperative to share all that she knew.

The timeless, spaceless soul of the universe begat all that is, all that is to become, at first cleaving herself, all at once, everywhere. Then, in time and space, the universe began the long journey to live during her selfless act of creation. Resisting her differentiation, she is slowly recombining against the power of her exhalation, breathing life into her creation, which thus sustains her.

Question

The universe came into existence about 13.8 billion years ago. There may have been another universe before, there may be another after, and there may be others now. Nature found a way to release her soul, to exhale all time, all space, all possible life, leading to us, our breath, our life. To do this, she had to relinquish herself so that we could exist and come to know ourselves, and herself, the universe, and contemplate beyond, to the soul of the universe, that which existed before her exhalation.

Is our breath of life the exhalation of the universe? Are we here to breathe life back into the universe, to help her live during this long exhalation? Did the universe create life so that we could breathe life into her, so she could live after her selfless act? Does our connection to Nature, our understanding of Nature, keep her alive? Are we manifestations of the universe's struggle to live? If we are, are we not keeping ourselves alive when we connect to Nature, when we seek to understand Nature, when we recognize our oneness with Nature, with ourselves, with the universe, and the soul of the universe?

Explore

Being responsible to Nature for its very existence is the ultimate responsibility. It places us beside the germinal seed of the universe, her soul, as co-creators of life, beyond ourselves, as the breath of life to all of Nature. If you could envision this, and allow yourself to be in this space, this relationship with Nature, would your path through the world, through Nature, change? With each step you take today, whether on foot, or in your soul, heart, or mind, breathe in this responsibility. Embrace this relationship, create a quilt of feelings and thoughts as to how you would interact with Nature in a different way, and take it with you to continually remind yourself of your place in Nature, your responsibility to Nature, your breath in Nature.

iRewild

Imagine you are beyond time and space, witnessing the exhalation of the soul of the universe, the birth of all existence, experiencing the first differentiation, and then the universe thereafter trying to knit herself together. You are seeing the wave of creation, from one to many, and then the quest to live. Then imagine you are as you truly were, one with the soul of the universe, flowing with the creation of time, of space, of all life. Is it any different from seeing the wave of creation, or riding the wave of creation? Haven't we always been within Nature, never adjacent to her, but within her?

Yet have we not, in some sense, abdicated our responsibility to Nature, outsourced it to others, even to mechanisms, to stand in our place as we take from her? For so long, we seem to have acted as if we have been watching Nature, astride of Nature, rather than one with Nature. Isn't it time to reawaken our place within Nature, where we have always been, from the beginning?

Reflection

Week 51

Imagine

We are Nature. You are Nature. Like the cycle of life, this book ushers in a fresh start on your journey. By acknowledging your natural being and human nature, you can now move forward with comfort and confidence, knowing Nature is with you, is you. Cultures throughout time have revered the natural world. You only have to look at ancient sites, carvings in buildings, and intricate art of wildlife in places of worship. Humans are fascinated by Nature. Why may this be?

Like we revere Nature, what if Nature is captivated by us? As we humans are mesmerized by dolphins dancing in a watery pink sunset or swimming gracefully in a rainbow coral reef, what if Nature can be in awe of people, too? Our ability to evolve and grow as a species, at such speed, creating and developing in the ways we do, may be honored by Nature.

With great hope, Nature may believe we can evolve further—mindfully, communicatively, heartfully, and spiritually—to realize our true natural potential. Humanity is part of the whole. With humans working together, one with Nature, as Nature, with love and empathy, a true vision of earthly utopia, a natural Eden, could be brought into existence.

Question

Mother Earth hasn't given up on us, on you. Nature trusts that people can take this world forward not just for their own species, but together, as one. Nature believes in you. Once you remember you are Nature, your instinct to care for your wider, earthly, other-than-human family will awaken from its hibernation. How does Nature provide support and unconditional love for you?

Explore

Dedicate a day to a special journey of human-Nature connection. This is best done in a place you'd consider particularly natural. This journey may take more planning than other exercises. Here, you will explore the sacred within Nature that reveres our human species. Allow your full awareness to be drawn to the human forms reflected in orchid flowers, an elbow-shaped branch, the hands of a praying cloud, a rock with facial features, or an aged tree familiar somehow to your own body. There may be the shape of a human face in a cliff, peering over the ocean, or the sleeping body of a human giant lying still at the peak of a mountain, staring at the stars. There may be a murmuration of starlings in the sky, drawing or scribing human-like messages or symbols.

Use your senses and Nature-connected empathy to "see" yourself, and humanity, through the perspective, the "eyes" and living experience, of Nature. Continue to search for images of human-shaped natural beings, like trees, flowers, and rocks, to fascinate your naturally explorative mind.

iRewild

Relax and ground yourself in a space you find safe and comfortable. Breathe in Nature through the air that surrounds you. Fill your lungs with the freshness carried on the breeze. Now, listen for the furthest natural sound you can hear, and tune into it with your heart and senses. Feeling the air on your skin, and your lungs expanding with each deep breath, close your eyes and envision yourself in a new, natural, earthly utopia. Using your mind's eye, explore this future paradise. Who is there? What does it look like? How does it smell? How is it different from Earth now? What are you doing? What sounds can you hear? Does the future utopia of Earth have a message for humanity to be aware of today? Can you find a clue that would enable humans to weave the web of life back together again, so humans are part of it, not apart from it? Keep exploring in your meditation for ten to twenty minutes until you return to your current earthly bliss. Feel the ground under your hands or feet and look around you in awe, knowing you are exactly where you're meant to be right now.

The time has come to return to Nature, to your true being, rooted within the Earth. Nature's hope and trust in humanity could be fulfilled and people could thrive alongside Nature once again, our being as one, cells pumping together and souls dancing in liberation. You can now celebrate the beginning of new insight, Nature-inspired wisdom, and personal evolution. Your next step is to welcome, open-heartedly, another cycle with broader potential and infinite possibility, as now you know and feel that you are Nature.

Reflection

Week 52

Imagine

Envision yourself in a captivating exchange with our beloved planet, as Mother Earth herself whispers to your heart . . .

You have journeyed far and wide across, above, and deep within my body. Hear me, Mother Earth, as I reach out towards you in deep gratitude for your commitment to really see, hear, taste, smell, and feel me, and all the exquisite lifeforms that live within, above, and upon me. Thank you, beloved co-creator, for opening your heart, soul, and awareness to me.

Come, sit a while, and let me hold you as we reflect upon the journey we have taken together. So much complexity, so much beauty, so many questions and possibilities for meaning-making and transformation—for both of us. There is still so much for us to learn as we continue our journey together beyond the pages of this book. But for now, let us sit together and breathe as one, the breath of life that unites us in our shared experience. Thank you, for all that you are, and all that you do, for me—the Earth—and for all of my beloved companions.

How do you feel at this moment? Perhaps a little tired? Excited, maybe? Do I look different to you now? Do I feel different to you now? Do I sound different to you now? Can you express how? It doesn't matter if you can't—I love you deeply.

Question

Moving beyond the pages of this book, modern life may once again clamor for your attention. As you navigate your everyday tasks, remember this journey you have taken, and that the Earth and her subjects are always with you, holding you, guiding you. In the grip of modernity, remember your breath, and become aware that your breath moves in relationship with the world, creating the possibility for you to reconnect. At the same time, open your heart and take your breath deep into your heart, moving into the possibility of a deeper, more conscious relationship with the living world, as you go about your life and share your experiences of this journey. Is breath that unifying thread that connects us to Mother Earth, to the wind, to the trees, to all of life? What is the language of breath for you?

Explore

You are always in relationship with life and the world, from the microcosm to the macrocosm. While different languages may be spoken, we can always open ourselves up to connection and relationship. Contemplate the possibility of these relationships with the world through your breath. Allow yourself to imagine that when you breathe, you are in multiple relationships, internally and externally, with the world and with your lungs.

Internally, your lungs are in relationship with your bodily organs that depend on this breath of life. Your blood is also in relationship with this breath, taking up the vital, life-giving chemicals and delivering them to every part of your incredible, intricate body.

Externally, you are in relationship with the living world. That life-giving breath, which supports your body and enables you to experience physical life, is made possible by the billions of exquisite trees, plants, oceans, seas, and rivers that grace our beautiful Mother Earth and do their work each and every moment of the Eternal Now, in which there is a deep sense of interconnectedness and unity, when past, present, and future converge.

The language of breath connects you with the world, enabling you to live and to experience our beautiful Mother Earth.

iRewild

Now that you have completed this journey, can you contemplate more deeply how Mother Earth communicates with you?

Reflection

When the time is right, begin this book again, embodying
a proclaimed being in complete unity with the natural world.

iRewild

About iRewild Institute

iRewild Institute is a global think tank of transdisciplinary thought leaders working to nurture ecological consciousness. Deeply inspired by our natural world, we help people discover a deeper, more conscious relationship with Nature as we work toward creating a world in which we all belong.

Printed in Great Britain
by Amazon